Author:
Mel Feigen, M.A.

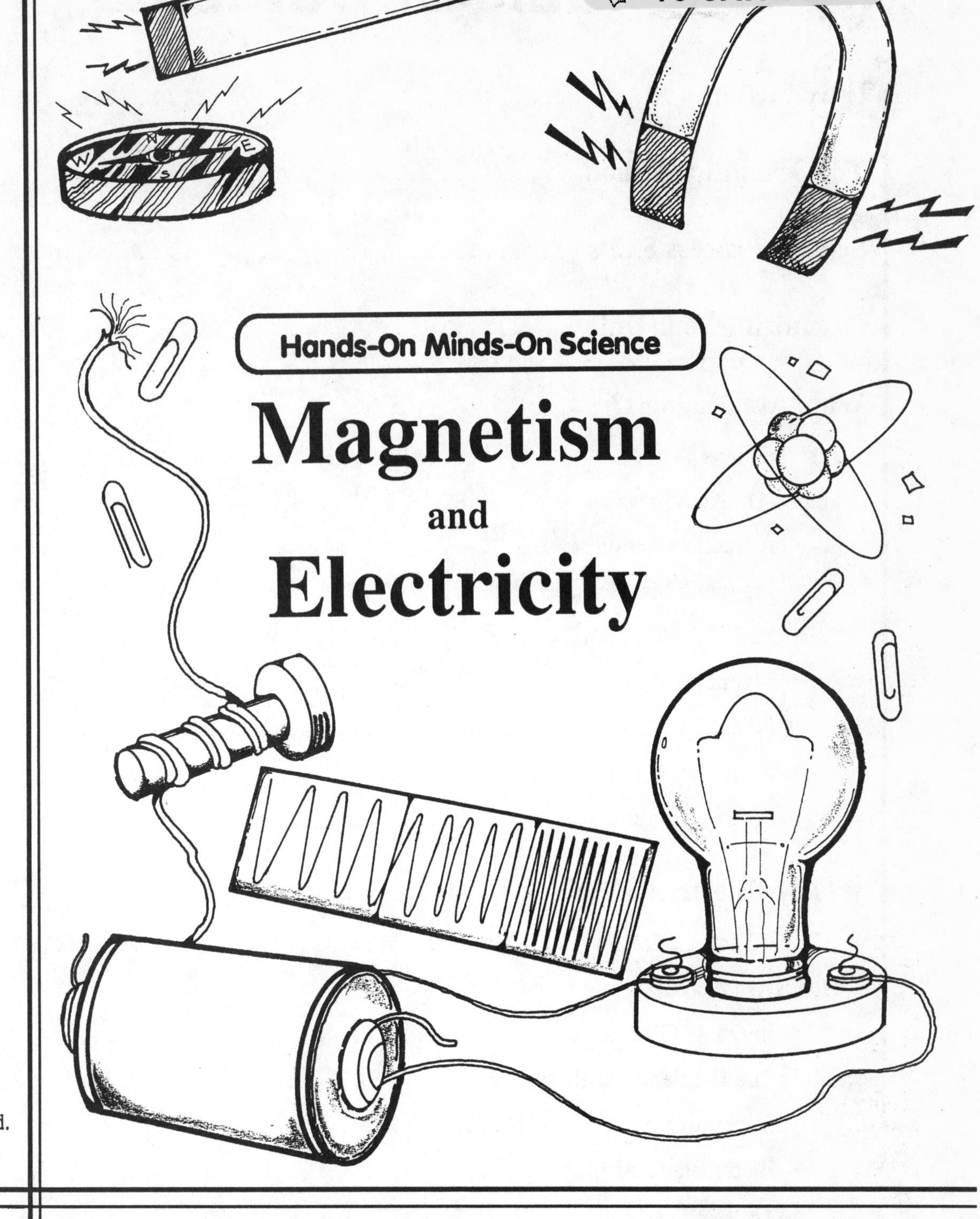

Magnetism and Electricity

Illustrator:
Sue Fullam

Editors:
Evan D. Forbes, M.S. Ed.
Walter Kelly, M.A.

Senior Editor:
Sharon Coan, M.S. Ed.

Art Director:
Darlene Spivak

Product Manager:
Phil Garcia

Imaging:
Rick Chacón

Research:
Bobbie Johnson

Publishers:
Rachelle Cracchiolo, M.S. Ed.
Mary Dupuy Smith, M.S. Ed.

Teacher Created Materials, Inc.
P.O. Box 1040
Huntington Beach, CA 92647

Made in U.S.A.

ISBN-1-55734-643-7

Table of Contents

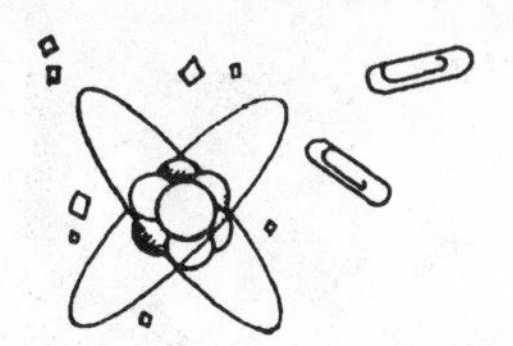

Table of Contents *(cont.)*

Introduction

What Is Science?

What is science to young children? Is it something that they know is a part of their world? Is it a textbook in the classroom? Is it a tadpole changing into a frog? Is it a sprouting seed, a rainy day, a boiling pot, a turning wheel, a pretty rock, or a moonlit sky? Is science fun and filled with wonder and meaning? What does science mean to children?

Science offers you and your eager students opportunities to explore the world around you and to make connections between the things you experience. The world becomes your classroom, and you, the teacher, a guide.

Science can, and should, fill children with wonder. It should cause them to be filled with questions and the desire to discover the answers to their questions. And, once they have discovered answers, they should be actively seeking new questions to answer.

The books in this series give you and the students in your classroom the opportunity to learn from the whole of your experience—the sights, sounds, smells, tastes, and touches, as well as what you read, write about, and do. This whole-science approach allows you to experience and understand your world as you explore science concepts and skills together.

Electricity, Magnets, and Electromagnets—What Are They?

Electricity and magnetism are energy forms that people have known about and studied for thousands of years. Although these energy forms have existed for so long, they have only had practical use in society for the last several hundred years. Neither source of energy can be seen, heard, or smelled, yet we know they exist. Prior to the use of these energies, people used other forms of energy to help with their work. For example, wood was burned to keep houses warm and cook food. Water and wind power were harnessed to grind grain and pump water.

During the nineteenth century, electromagnetism was discovered when Hans Oersted realized electric current produced a magnetic field. The first practical electromagnet was built by Joseph Henry and put to use in the late 1820's. Today, electromagnets are used in doorbells, buzzers, as relays, and to lift heavy metals made of iron.

The Scientific Method

The scientific method is a creative and systematic process for proving or disproving a given question, following an observation. When scientists use the scientific method, a basic set of guiding principles and procedures is followed in order to obtain new knowledge about our universe. This method will be described in the paragraphs that follow.

It is easy to teach the scientific method! Just follow these simple steps:

Make an **OBSERVATION**.

The teacher presents a situation, gives a demonstration, or reads background material that interests students and prompts them to ask questions. Or students can make observations and generate questions on their own as they study a topic.

Example: A working light.

Select a **QUESTION** to investigate.

In order for students to select a question for a scientific investigation, they will have to consider the materials they have or can get, as well as the resources (books, magazines, people, etc.) actually available to them. You can help them make an inventory of their materials and resources, either individually or as a group.

Tell students that in order to successfully investigate the questions they have selected, they must be very clear about what they are asking. Discuss effective questions with your students. Depending upon their level, simplify the question or make it more specific.

Example: What happens when a circuit is completed?

Make a **PREDICTION** *(Hypothesis)*.

Explain to students that a hypothesis is a good guess about what the answer to a question will probably be. But they do not want to make just any arbitrary guess. Encourage students to predict what they think will happen and why.

In order to formulate a hypothesis, students may have to gather more information through research.

Have students practice making hypotheses with questions you give them. Tell them to pretend they have already done their research. You want them to write each hypothesis so it follows these rules:

1. It is to the point.
2. It tells what will happen, based on what the question asks.
3. It follows the subject/verb relationship of the question.

Example: I think when a circuit is completed, the flow of electricity will cause the light to turn on.

The Scientific Method *(cont.)*

Develop a **PROCEDURE** to test the hypothesis.

The first thing students must do in developing a procedure (the test plan) is to determine the materials they will need.

They must state exactly what needs to be done in step-by-step order. If they do not place their directions in the right order, or if they leave out a step, it becomes difficult for someone else to follow their directions. A scientist never knows when other scientists will want to try the same experiment to see if they end up with the same results!

Example: You will build a model of a simple circuit and then test it to see if it works.

Record the **RESULTS** of the investigation in written and picture form.

The results (data collected) of a scientific investigation are usually expressed two ways—in written form and in picture form. Both are summary statements. The written form reports the results with words. The picture form (often a chart or graph) reports the results so the information can be understood at a glance.

Example: The results of the investigation can be recorded on a data-capture sheet provided (page 41).

State a **CONCLUSION** that tells what the results of the investigation mean.

The conclusion is a statement which tells the outcome of the investigation. It is drawn after the student has studied the results of the experiment, and it interprets the results in relation to the stated hypothesis. A conclusion statement may read something like either of the following: "The results show that the hypothesis is supported," or "The results show that the hypothesis is *not* supported." Then restate the hypothesis if it was supported or revise it if it was not supported.

Example: The hypothesis that stated, "when the circuit is completed, the flow of electricity will cause the light to turn on," is supported (or *not supported).*

Record **QUESTIONS, OBSERVATIONS,** and **SUGGESTIONS** for future investigations.

Students should be encouraged to reflect on the investigations that they complete. These reflections, like those of professional scientists, may produce questions that will lead to further investigations.

Example: Are all circuits the same?

Science-Process Skills

Even the youngest students blossom in their ability to make sense out of their world and succeed in scientific investigations when they learn and use the science-process skills. These are the tools that help children think and act like professional scientists.

The first five process skills on the list below are the ones that should be emphasized with young children, but all of the skills will be utilized by anyone who is involved in scientific study.

Observing

It is through the process of observation that all information is acquired. That makes this skill the most fundamental of all the process skills. Children have been making observations all their lives, but they need to be made aware of how they can use their senses and prior knowledge to gain as much information as possible from each experience. Teachers can develop this skill in children by asking questions and making statements that encourage precise observations.

Communicating

Humans have developed the ability to use language and symbols which allow them to communicate not only in the "here and now" but over time and space as well. The accumulation of knowledge in science, as in other fields, is due to this process skill. Even young children should be able to understand the importance of researching others' communications about science and the importance of communicating their own findings in ways that are understandable and useful to others. The magnetism and electricity journal and the data-capture sheets used in this book are two ways to develop this skill.

Comparing

Once observation skills are heightened, students should begin to notice the relationships between things that they are observing. Comparing means noticing the similarities and differences. By asking how things are alike and different or which is smaller or larger, teachers will encourage children to develop their comparison skills.

Ordering

Other relationships that students should be encouraged to observe are the linear patterns of seriation (order along a continuum: e.g., rough to smooth, large to small, bright to dim, few to many) and sequence (order along a time line or cycle). By making graphs, time lines, cyclical and sequence drawings, and by putting many objects in order by a variety of properties, students will grow in their abilities to make precise observations about the order of nature.

Categorizing

When students group or classify objects or events according to logical rationale, they are using the process skill of categorizing. Students begin to use this skill when they group by a single property such as color. As they develop this skill they will be attending to multiple properties in order to make categorizations; the animal classification system, for example, is one system students can categorize.

Science-Process Skills *(cont.)*

Relating

Relating, which is one of the higher-level process skills, requires student scientists to notice how objects and phenomena interact with one another and the changes caused by these interactions. An obvious example of this is the study of chemical reactions.

Inferring

Not all phenomena are directly observable, because they are out of humankind's reach in terms of time, scale, and space. Some scientific knowledge must be logically inferred based on the data that is available. Much of the work of paleontologists, astronomers, and those studying the structure of matter is done by inference.

Applying

Even very young, budding scientists should begin to understand that people have used scientific knowledge in practical ways to change and improve the way we live. It is at this application level that science becomes meaningful for many students.

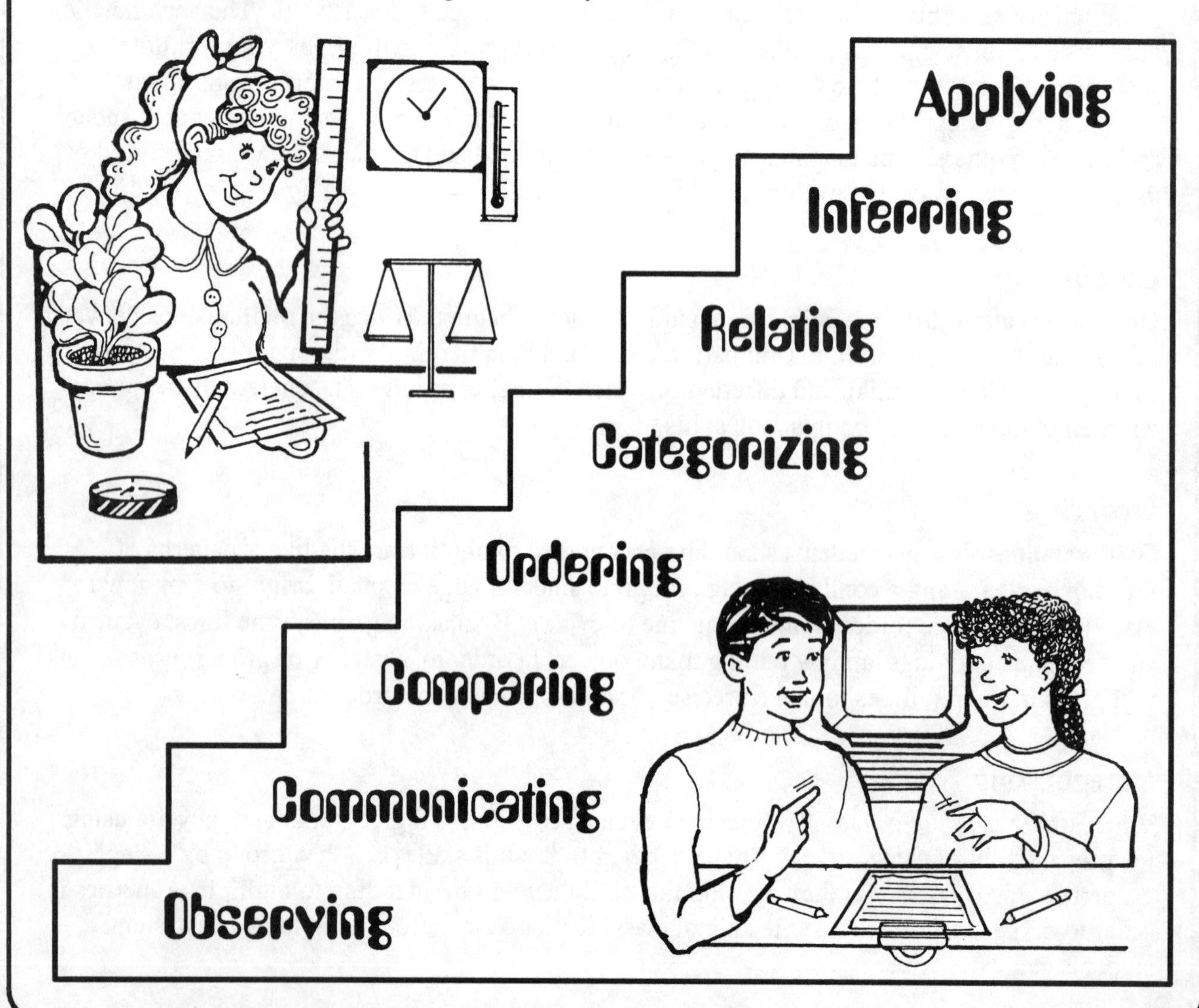

Organizing Your Unit

Designing a Science Lesson

In addition to the lessons presented in this unit, you will want to add lessons of your own, lessons that reflect the unique environment in which you live, as well as the interests of your students. When designing new lessons or revising old ones, try to include the following elements in your planning:

Question

Pose a question to your students that will guide them in the direction of the experience you wish to perform. Encourage all answers, but you want to lead the students towards the experiment you are going to be doing. Remember, there must be an observation before there can be a question. (Refer to The Scientific Method, pages 5-6.)

Setting the Stage

Prepare your students for the lesson. Brainstorm to find out what students already know. Have children review books to discover what is already known about the subject. Invite them to share what they have learned.

Materials Needed for Each Group or Individual

List the materials each group or individual will need for the investigation. Include a data-capture sheet when appropriate.

Procedure

Make sure students know the steps to take to complete the activity. Whenever possible, ask them to determine the procedure. Make use of assigned roles in group work. Create (or have your students create) a data-capture sheet. Ask yourself, "How will my students record and report what they have discovered? Will they tally, measure, draw, or make a checklist? Will they make a graph? Will they need to preserve specimens?" Let students record results orally, using a video or audio tape recorder. For written recording, encourage students to use a variety of paper supplies such as poster board or index cards. It is also important for students to keep a journal of their investigation activities. Journals can be made of lined and unlined paper. Students can design their own covers. The pages can be stapled or be put together with brads or spiral binding.

Extensions

Continue the success of the lesson. Consider which related skills or information you can tie into the lesson, like math, language arts skills, or something being learned in social studies. Make curriculum connections frequently and involve the students in making these connections. Extend the activity, whenever possible, to home investigations.

Closure

Encourage students to think about what they have learned and how the information connects to their own lives. Prepare journals using "Magnetism and Electricity Journal" directions on page 80. Provide an ample supply of blank and lined pages for students to use as they complete the "Closure" activities. Allow time for students to record their thoughts and pictures in their journals.

The Big Why

The explanation behind the experience is provided.

Organizing Your Unit *(cont.)*

Structuring Student Groups for Scientific Investigations

Using cooperative learning strategies in conjunction with hands-on and discovery learning methods will benefit all of the students taking part in the investigation.

Cooperative Learning Strategies

1. In cooperative learning all group members need to work together to accomplish the task.
2. Cooperative learning groups should be heterogeneous.
3. Cooperative learning activities need to be designed so that each student contributes to the group and individual group members can be assessed on their performance.
4. Cooperative learning teams need to know the social as well as the academic objectives of a lesson.

Cooperative Learning Groups

Groups can be determined many ways for the scientific investigations in your class. Here is one way of forming groups that has proven to be successful in primary classrooms.

- **Project Leader—**scientist in charge of reading directions and setting up equipment
- **Physicist—**scientist in charge of carrying out directions (can be more than one student)
- **Stenographer—**scientist in charge of recording all of the information
- **Transcriber—**scientist who translates notes and communicates findings

If the groups remain the same for more than one investigation, require each group to vary the people chosen for each job. All group members should get a chance to try each job at least once.

Using Centers for Scientific Investigations

Set up stations for each investigation. To accommodate several groups at a time, stations may be duplicated for the same investigation. Each station should contain directions for the activity, all necessary materials (or a list of materials for investigators to gather), a list of words (a word bank) which students may need for writing and speaking about the experience, and any data-capture sheets or needed materials for recording and reporting data and findings.

Station-to-Station Activities are on pages 67-78. Model and demonstrate each of the activities for the whole group. Have directions at each station. During the modeling session, have a student read the directions aloud while the teacher carries out the activity. When all students understand what they must do, let small groups conduct the investigations at the centers. You may wish to have a few groups working at the centers while others are occupied with other activities. In this case, you will want to set up a rotation schedule so all groups have a chance to work at the centers.

Assign each team to a station, and after they complete the task described, help them rotate in a clockwise order to the other stations. If some groups finish earlier than others, be prepared with another unit-related activity to keep students focused on main concepts.

After all rotations have been made by all groups, come together as a class to discuss what was learned.

Just the Facts

Observe a magnet. Is there anything around it you can see? Now put your hand next to the magnet. Do you feel anything? You were unable to see anything close to the magnet and your hand could feel nothing next to the magnet. But there is a force there that is invisible and "untouchable." We call this force magnetism. This is the invisible force that makes a compass needle point north.

Lodestone or magnetic rock was known by ancient peoples. They knew that if a lodestone was hung freely it would point in a northerly direction. Some unknown person placed a magnetized needle on a pivot, and the first magnetic compass was invented.

Peregrinus was the person who named the ends of a magnet. He named the end that points north the North Pole and the end that points south the South Pole. But no one knew why the ends of a magnet pointed north and south.

Finally in 1600, Dr. William Gilbert, an Englishman, showed that the earth itself is a giant magnet.

Some facts about magnets:

1. Objects attracted by a magnet contain iron or steel.
2. Like magnetic poles repel.
3. Opposite magnetic poles attract.
4. Each magnet has a magnetic field around it. It is strongest close to the magnet. For an object to be affected by the magnet, it must be within the magnetic field.
5. If you cut a magnet in half, you will end up with a new magnet. The two new magnets will both have north and south poles. Nobody knows the reason for this strange effect.
6. Magnets can make ordinary steel objects into magnets themselves. For example, a nail can be made magnetic by rubbing it with a magnet.
7. Place a piece of cardboard between a magnet and an iron object, and the magnetic force passes through the cardboard and still attracts the iron object.

Attraction Action

Question

What is attracted to a magnet?

Setting the Stage

- Hold a discussion with students about magnetism to find out what they already know.
- Set up a display in the classroom showing different magnets and how they are used.

Materials Needed for Each Group

- bar magnet
- several magnetic objects (paper clips, nails, iron filings, etc.)
- several nonmagnetic objects (plastic, paper, coins, etc.)
- pen or pencil
- data-capture sheet (page 13), one per student

Procedure *(Student Instructions)*

1. Using a bar magnet, select one object at a time to see if it is magnetic.
2. Record your findings on the data-capture sheet provided.
3. Once you have tested all of the objects provided, be creative and find five more objects to test for magnetic properties.
4. Repeat step 2.

Extensions

- Have students compare the magnetic attraction of a bar magnet with that of other types of magnets, using the same objects in the above experiment.
- With a ruler, have the students measure the distance between the objects and the magnet when the attraction first began.

Closure

In their electricity and magnet journals, have students create a magnetic sculpture using several small magnets and iron or steel pieces and then draw a picture of their creation.

The Big Why

Magnets attract materials that are magnetic, such as iron, steel, cobalt, and nickel. Why metals are magnetic or nonmagnetic is still unknown.

Attraction Action *(cont.)*

Record your findings on the chart and answer the questions on either the back or a separate sheet of paper.

TYPE OF OBJECT	ATTRACTED TO THE MAGNET	NOT ATTRACTED TO THE MAGNET

1. Do the magnets attract all metal objects?
2. What part of the magnet has the object attracted?

Magnetic Muscle

Question

How can you test the strength of a magnet?

Setting the Stage

- Review with students the concept of magnetism.
- Have students brainstorm ways to test the strength of nonmagnetic materials such as paper, rubber bands, and thread. Some of the ideas for testing paper, rubber bands, and thread may be adapted to test the strength of magnets.

Materials Needed for Each Group

- several magnets of different sizes and shapes
- 20 small nails
- 20 metal paper clips
- data-capture sheet (page 15), one per student

Procedure *(Student Instructions)*

1. With the 20 nails make a pile on a desk or other hard surface.
2. Choose one magnet and place it in the pile of nails.
3. Slowly lift it out of the pile and record the number of nails it picks up on the data-capture sheet provided.
4. Repeat steps 2-3 with the remaining magnets.
5. Replace the nails with the 20 paper clips and repeat steps 2-4. Compare your results from the first experience.

Extensions

- Have students come up with other ideas for testing the strength of magnets and put those ideas into practice.
- Have students place a magnet at one end of a ruler and a nail at the other. Slowly move the magnet closer to the nail. Record the distance when the nail is first attracted to the magnet. Repeat this with a paper clip.

Closure

In their magnetism and electricity journals, have students write a paragraph bringing to life (personifying) their magnet holding on to some object.

The Big Why

The number of nails or paper clips a magnet picks up will help to determine its relative strength.

Magnetic Muscle *(cont.)*

Fill in the chart and answer the questions.

Type of Magnet	Number of Nails Held?	Number of Paper Clips Held?

1. Which magnet seemed to be the strongest?

____

____

____

2. Did the size of the magnet affect its strength?

____

____

____

3. Were the magnets able to pick up more nails than paper clips? Why?

____

____

____

Power Ends

Question

What part of a magnet has the strongest attraction?

Setting the Stage

- Have students define the term *magnetic pole.*
- Ask students why they would think one part of a magnet would be stronger than another.
- Ask students if the shape of a magnet has any effect on where it is strongest.

Materials Needed for Each Group

- bar magnet
- spring scale
- large paper clip
- clear tape
- ruler
- pencil
- data-capture sheet (page 17), one per student

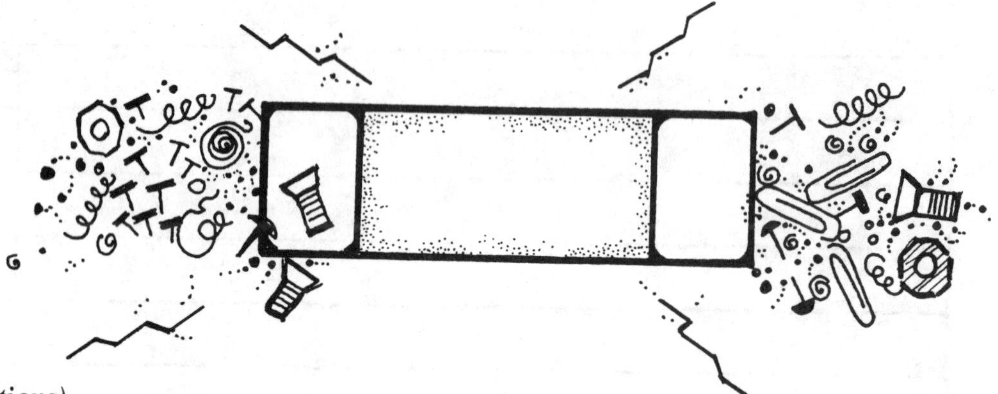

Procedure *(Student Instructions)*

1. Tape the bar magnet onto a desk or table.
2. Using a ruler and a pencil, make ½" (1.25 cm) marks along the entire length of the magnet.
3. Hook the paper clip onto the end of the spring scale.
4. Starting at one end of the bar magnet and moving along each ½" (1.25 cm) mark, test the pull required to lift the paper clip from the magnet.
5. Chart your results on the data-capture sheet, noting the amount of pull and the distance from the end of the magnet.

Extensions

- Have students hold a bar magnet in one hand. In the other hand have them hold a paper clip over the magnet. Have them drop the paper clip and notice where it is attracted.
- Have students lay a bunch of nails or paper clips on a desk. Then place a bar magnet over the pile horizontally and observe where the majority of nails or paper clips are attached.

Closure

In their magnetism and electricity journals, have students write about their experiences with magnetic poles.

The Big Why

Magnets are strongest at their poles.

Power Ends *(cont.)*

Record the reading from your spring scale for each 1/2" (1.25 cm) mark and then answer the questions.

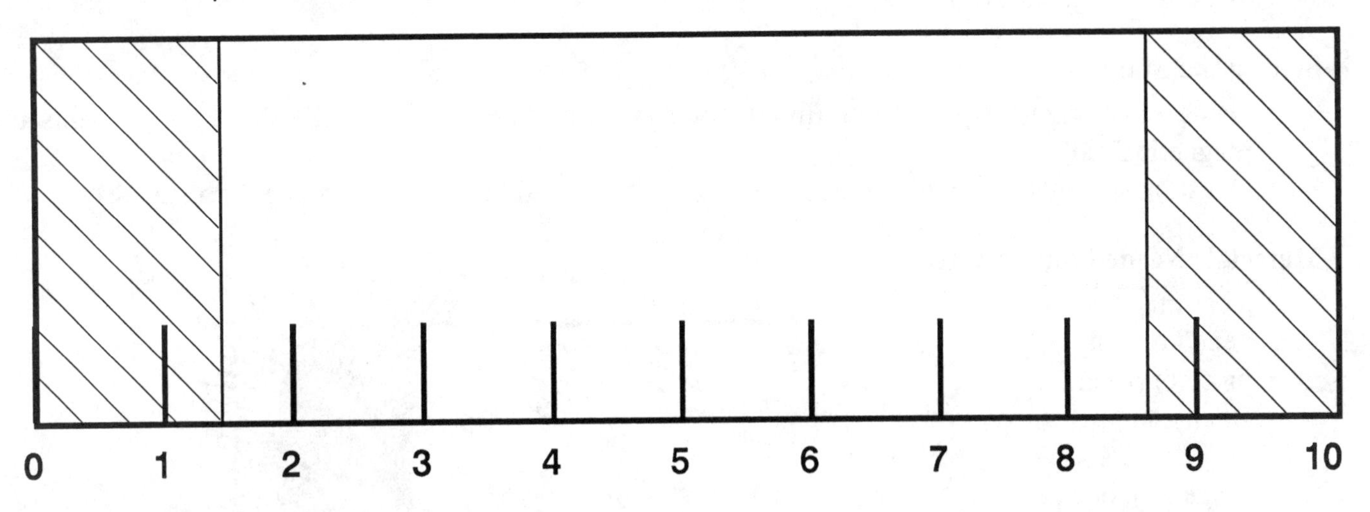

1. Is there a noticeable change in the magnet's strength as you go from one side to the other? Explain.

____

____

____

____

2. Where is the strongest part of the magnet?

____

____

____

____

3. Do you think all magnets are strongest at their poles and weaker as you move away? Explain.

____

____

____

____

Magnetic Map

Question

What is a magnetic field?

Setting the Stage

- Using a bar magnet, a pile of iron filings, and an overhead projector, visually show your students a magnetic field.
- Question students as to why the iron filings can be affected by a magnet from a distance away.

Materials Needed for Each Group

- a bar magnet
- pencil or pen
- piece of paper
- small compass
- data-capture sheet (page 19), one per student

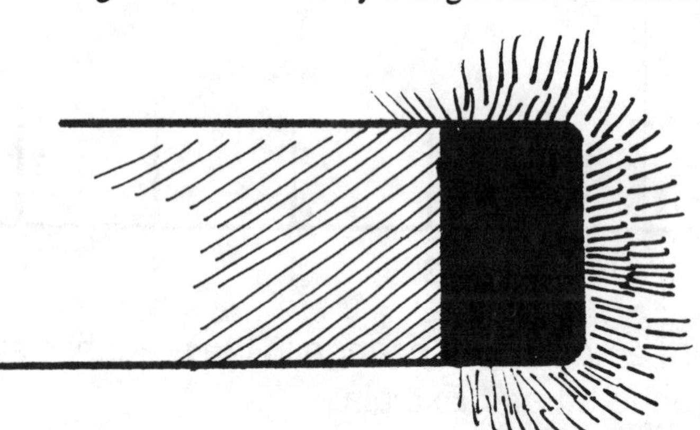

Procedure *(Student Instructions)*

1. Place the bar magnet in the middle of the piece of paper and trace around it to mark its position.
2. Put the compass on the paper near the magnet. Draw an arrow between the compass and the magnet showing the farthest point where the compass needle is affected by the magnetic field.
3. Repeat step two several times all around the magnet.
4. Once you have completed step three, you will be able to see where the magnetic field exists, where it is strongest, and where it is weakest.
5. With the information gained from this experience, answer the questions on your data-capture sheet.

Extensions

- Have students repeat the experiment with a stronger magnet to see if the magnetic field gets any larger.
- Using magnetic objects, have the students place them on the rim, inside the rim, and outside the rim of the magnetic field and record their reactions.
- Have students create their own discrepant event (an event inconsistent with science), using a magnet, paper clip, and a piece of thread.

Closure

In their magnetism and electricity journals, have students write a paragraph about what it would be like to be a magnetic object being pulled through a magnetic field towards a magnet.

The Big Why

The magnetic field is the force that exists around the magnet. You can not actually see this field, but by completing this experience, you will understand where the magnetic field is strongest and weakest.

Magnetic Map *(cont.)*

Answer the following questions.

1. What is a magnetic field?

 ____

 ____

 ____

2. Where around the bar magnet is the magnetic field the strongest?

 ____

 ____

 ____

3. Why does the compass needle point towards the magnet when it is within the magnetic field?

 ____

 ____

 ____

4. What happens to the compass needle when it is no longer affected by the magnetic field?

 ____

 ____

 ____

Poles Apart

Questions

What happens when you try to put together magnets with like poles? With unlike poles?

Setting the Stage

- Have students determine the north and south poles of a magnet with the use of a compass.
- Explain to students that magnetic north is not the same as geographic north.

Materials Needed for Each Group

- several lengths of thread
- two bar magnets
- two small rubber bands
- compass
- pen or pencil
- data-capture sheet (page 21), one per student

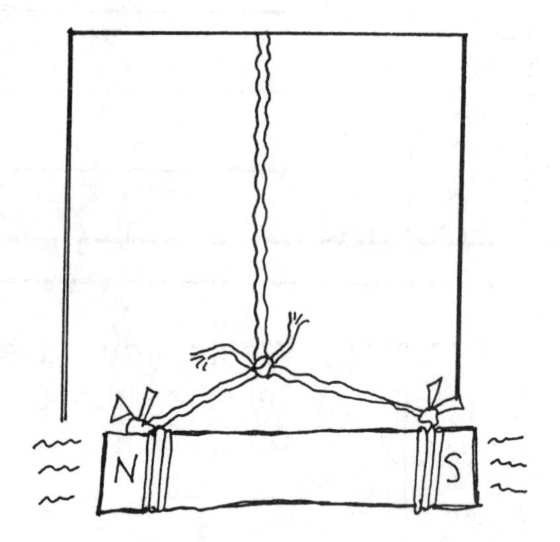

Procedure *(Student Instructions)*

1. Take one bar magnet and secure a rubber band to each end.
2. Using one length of string, tie the ends of the string to each of the rubber bands.
3. Then, with a second length of string, tie it to the middle of the string holding the bar magnet.
4. Suspend the magnet so it can swing freely. Note: Wait for the magnet to stop moving before beginning the experience.

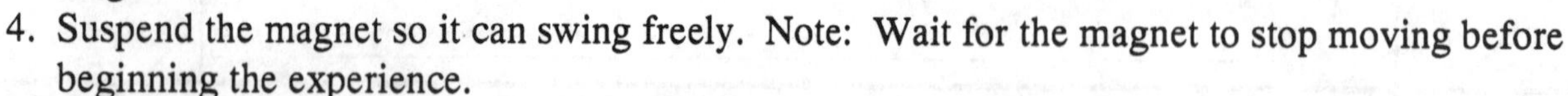

5. Using a compass, have someone line up the bar magnet, to learn which end is the north pole and which end is the south pole.
6. Then have someone place the north pole of the second bar magnet near the north pole of the suspended magnet. Observe what happens and record observations on the data-capture sheet.
7. Then have someone place the south pole of the second bar magnet near the south pole of the suspended magnet. Observe what happens and record observations on the data-capture sheet.
8. Finally, have someone place the south pole of the second bar magnet near the north pole of the suspended magnet. Observe what happens and record observations on the data-capture sheet.

Extensions

- Have students dip magnets into a pile of iron filings and place like poles near each other, then unlike poles near each other, and observe what happens.
- Have students try to print their names using bar magnets to construct the letters.

Closure

In their magnetism and electricity journals, have students come up with their own experiences showing how like poles repel and unlike poles attract.

The Big Why

Like poles repel, but unlike poles attract.

Poles Apart *(cont.)*

Fill in the information required.

1. Draw a picture and describe with sentences what you observed when both north poles were put near each other.

2. Draw a picture and describe with sentences what you observed when both south poles were put near each other.

3. Draw a picture and describe with sentences what you observed when both opposite poles were put near each other.

Neighborly Nail

Question

Is it possible to turn an ordinary nail into a magnet?

Setting the Stage

- Review with students the properties of magnets.
- Ask students if they have ever seen something besides a magnet pick up an object in a magnetic way.
- Have students make a class list of their ideas.
- Have students brainstorm ways on how to make objects magnetic.

Materials Needed for Each Group

- large nail
- bar magnet
- iron fillings
- small nails or paper clips
- data-capture sheet (page 23), one per student

Procedure *(Student Instructions)*

1. With the large nail try to pick up some of the smaller nails or paper clips. Observe what happens and record on data-capture sheet.
2. Now, with one end of the bar magnet, stroke the nail 25 times in the same direction.
3. Try again to pick up the small nails or paper clips with the newly magnetized nail. Observe what happens and record on data-capture sheet.
4. Carefully throw the magnetized nail against a hard surface.
5. Try one more time to pick up the small nails or paper clips with the nail. Observe what happens and record on data-capture sheet.
6. Repeat this experience using iron fillings instead of nails or paper clips. Observe what happens and record on data-capture sheet.

Extensions

- Have students stroke the nail more than 25 times to see if the nail will become a stronger magnet.
- Have students try and magnetize a variety of other metal objects and repeat the above experience.
- Discuss with students why the nail became demagnetized when it was thrown against something hard.

Closure

In their magnetism and electricity journals, have students draw a picture of what the atoms of a magnetized and demagnetized nail look like.

The Big Why

Stroking the magnet along the nail causes the atoms in the nail to line up, becoming magnetized.

Neighborly Nail *(cont.)*

In the boxes draw what you saw in your experience.

Picking up paper clips on a nail.

Nonmagnetized Nail	Magnetic Nail	Demagnetized Nail

Picking up iron fillings on a nail.

Nonmagnetized Nail	Magnetic Nail	Demagnetized Nail

Floating Compass

Question

Using magnets, how can you make a simple compass?

Setting the Stage

- Discuss with students how magnetic poles are used in telling direction.
- Discuss with students the history of the compass.
- Have students do research on earliest users of the compass.

Materials Needed for Each Group

- a needle
- a bar magnet
- dish of water
- piece of paper larger than the dish of water
- piece of cork or styrofoam
- compass
- data-capture sheet (page 25), one per student

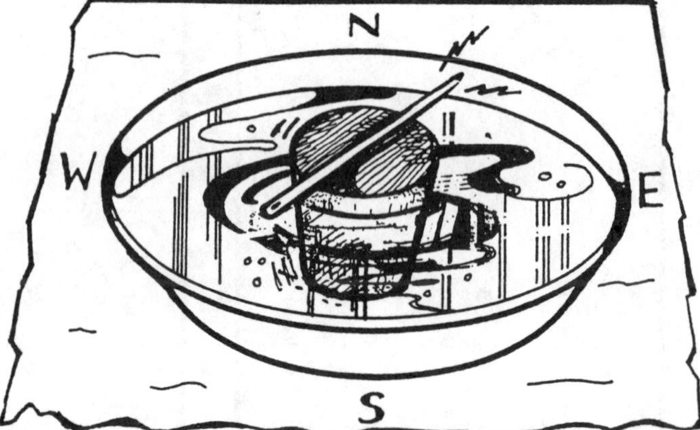

Procedure *(Student Instructions)*

1. Float a piece of cork or styrofoam in a dish of water.
2. Magnetize the needle by stroking it 25 times in the same direction, with one end of the magnet.
3. Test the needle with a compass to make sure it is magnetized.
4. Carefully set the needle on top of the cork or styrofoam and wait for it to stop moving.
5. On the piece of paper mark the top with an "N," mark the bottom with an "S," mark the left side with a "W," and mark the right side with an "E."
6. Place the dish of water on top of the paper and match up the needle with the "N." Your compass is complete.
7. To check if the compass works, turn the cork or styrofoam away from "N" and let go to see if the needle returns to "N."

Extensions

- Have students label their compasses with degrees and then try to find specific compass headings.
- Have students research why compass needles naturally point north-south instead of east-west.

Closure

In their magnetism and electricity journals, have students write a story of how they crossed the ocean using a simple compass.

The Big Why

Once the needle becomes magnetized, it will line up with the earth's magnetic north pole.

Floating Compass *(cont.)*

Answer the following questions.

1. Compare the compass you made to one you can buy in a store. What are the similarities and differences you see?

2. Is the compass you made easy and practical to use? Explain.

3. Does the needle of your simple compass point to magnetic north or geographic north? Are they the same place? Explain.

4. Draw a picture of your simple compass.

Just the Facts

Electricity is the most widely used source of power worldwide. Make a chart of all the different uses you can think of for electricity. Have others in the room add to your chart ideas. Electrical energy is an important part of our lives. Businesses, homes, and schools are all powered by electricity. The lights, radios, power tools, and televisions in our homes are all run by electricity.

Electricity can be generated from many sources.

1. **Power Stations**—Giant turbines are turned to generate the electricity that powers our homes. This electricity is carried to our homes by overhead lines or underground cables. Nuclear power may also be used to run these stations.
2. **Batteries**—Batteries use a chemical reaction to make electric current. Eventually all of the major automakers will have cars that run on batteries. Fueling up will someday consist of just plugging into an electrical outlet.
3. **Nature's Electricity**—Lightning is an important show of nature's electricity. Also certain animals and insects show electrical abilities. The electric eel and glowworm are examples of this.
4. **Solar cells—**They are often used for today's high-tech energy needs. These cells get energy from light which they turn into electrical energy.

There are two types of electricity—*static* and *current*. The type we use the most is current. This is electricity which flows continuously. Static electricity is caused by rubbing two things together. Static electricity jumps, and current electricity flows.

Your body uses electricity in its nervous system. Messages are sent from sensory nerves to the brain. Muscles are given messages to move by electrical signals from the brain. Electrical signals in the body travel about 250 miles/hour (400 km/hour).

Invisible Glue

Question

What causes static electricity?

Setting the Stage

- Have students share their ideas about electricity.
- Discuss with students the make-up of an atom—electrons, neutrons, and protons.
- Have students do research about static electricity and then share what they have learned with the class.

Materials Needed for Each Group

- several round balloons
- different pieces of material (e.g., wool, 100% cotton, 50% cotton, polyester, paper, plastic, etc.)
- stopwatch
- data-capture sheet (page 28), one per student

Procedure *(Student Instructions)*

1. Blow up a balloon and tie it off for each piece of cloth you are going to test.
2. Take one piece of material and rub it five times against a balloon.
3. Try to stick that balloon against a wall.
4. Using a stopwatch, time how long the balloon remains on the wall.
5. Chart the observations from your experience on the data-capture sheet provided.
6. Repeat steps 2-5 until all of the materials have been tested.

Extensions

- Have students repeat the experience by increasing or decreasing the number of times they rub the balloon and see if that affects the outcome.
- Have students repeat the experience by rubbing their heads with the balloons and then sticking the balloons against a wall.
- Discuss with students how lightning is caused by static electricity in the atmosphere.

Closure

In their magnetism and electricity journals, have students write a creative essay in which each is a jumping electron. Have them tell about their adventures and daring jumps.

The Big Why

Rubbing a balloon with certain materials causes the balloon to pick up additional electrons. This gives the balloon a negative charge. The wall is neutral, creating a positive charge when the balloon gets close to the wall, thus attracting the balloon to the wall. Unlike charges attract.

Invisible Glue *(cont.)*

Answer the following questions about your experience.

Type of Material	Number of Times Rubbed	Time on Wall

1. Why or why not did the balloons stick to the wall?

____

____

2. Did using different types of materials have an effect on why the balloons stuck to the wall? Explain.

____

____

3. Would rubbing the balloons more frequently allow them to stay attracted to the wall for a longer period of time?

____

____

4. What other factors, if any, would cause the balloons to shorten their stay on the wall?

____

____

5. Would a balloon stay on the wall if it did not have a charge?

____

____

The Unfriendly Balloon

Question

What happens when a balloon becomes charged with static electricity?

Setting The Stage

- Have students discuss their ideas of electricity.
- Show students several examples of how electricity is used (past and present).
- Show students a model of an atom. Explain the three different types of particles found in an atom. Then explain how an atom becomes electrically charged.

Materials Needed for Each Group

- several round balloons
- several lengths of string or thread
- one spray bottle
- a wool cloth
- pen or pencil
- colored markers or crayons
- data-capture sheet (page 30)

Procedure *(Student Instructions)*

1. Blow up and knot two balloons.
2. Tie a length of string or thread a 1' (30 cm) long to the end of each balloon.
3. Hold together the ends of the string or thread to which the balloons are attached. Observe what happens and record observations on the data-capture sheet.
4. Using a wool cloth, rub both balloons and put them together. Observe what happens and record observations on the data-capture sheet.
5. Lightly mist one of the balloons and put them together. Observe what happens and record observations on the data-capture sheet.
6. Lightly mist both balloons and put them together. Observe what happens and record observations on data-capture sheet.

Extensions

- Have students rub the balloon using different types of material to see if they can get a static charge.
- Using a stopwatch, have the students time how long the static charges last on the balloons.

Closure

In their electricity and magnet journals, have the students write a story about a statically charged balloon.

The Big Why

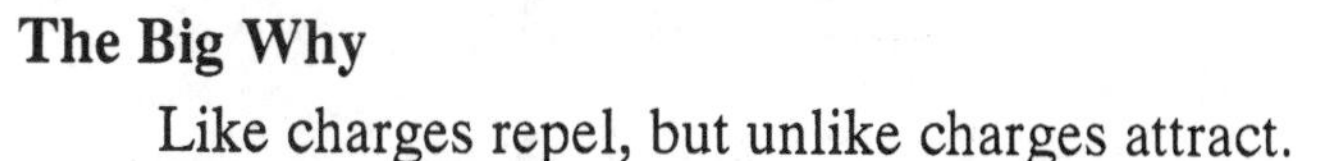

Like charges repel, but unlike charges attract.

The Unfriendly Balloon *(cont.)*

Fill in the information required.

1. Draw a picture and describe what happened to the two balloons when they were rubbed with a wool cloth and put together.

2. Draw a picture and describe what happened when one of the balloons was misted with water and put next to the balloon that remained the same.

3. Draw a picture and describe what happened when both balloons were misted with water and put next to each other.

Sticky Comb

Question

Will static electricity affect the flow of water?

Setting the Stage

- Review with students the reasons that static electricity exists.
- Have students create their own static charge at home. Have them rub their feet on a carpet and then touch something metal to create a static charge.
- Have students share their experiences with static charges.

Materials Needed for Each Group

- rubber comb
- piece of wool cloth
- several pieces of ripped paper
- working water tap
- data-capture sheet (page 32), one per student

Procedure *(Student Instructions)*

1. Rub a piece of wool cloth several times against your rubber comb.
2. Go over to a sink, turn the water so there is a gentle stream, and hold the comb near the stream of water.
3. Observe what is happening. Record on the data-capture sheet provided.
4. Repeat step 1, but this time put the comb near some ripped-up pieces of paper.
5. Observe what is happening. Record on the data-capture sheet provided.

Extensions

- Have students test other materials to see if they can create an attracting charge.
- Have students research static electricity to see if it is used as an energy source.

Closure

In their magnetism and electricity journals, have students write in their own words why the water was pulled towards the comb.

The Big Why

When the comb is rubbed with the wool cloth, it becomes negatively charged. Running water has a neutral charge, so when the comb is placed near the water the electrons in the water closest to the comb move away, causing the water to be positively charged. Therefore, the unlike charges attract the water to the comb.

Sticky Comb *(cont.)*

Draw a picture of what you are observing.

Draw a picture of the water bending toward the comb.

Draw a picture of the paper being attracted toward the comb.

Electrostatic Airplane

Question

How can two objects change from being attracted to each other, to repelling each other?

Setting the Stage

- Discuss with students the concept that like charges repel and unlike charges attract.
- Have students give as many examples as possible to show their understanding of this concept.

Materials Needed for Each Group

- plastic ruler or rod
- scissors
- piece of thin aluminum foil
- wool cloth
- stopwatch
- data-capture sheet (page 34), one per student

Procedure *(Student Instructions)*

1. Cut out and make a small airplane from the aluminum foil.
2. Using the wool cloth, rub the ruler several times to give it a negative charge.
3. Slowly bring the ruler close to the aluminum foil airplane. Observe what happens and record on your data-capture sheet.
4. With the ruler, hold the attached airplane high in the air. Observe what happens to the airplane and record on your data-capture sheet.
5. Repeat the experience and time the flight of your airplane.

Extensions

- Have students repeat the experience several times to see if they can control the direction of the airplane's flight.
- Have students do research on gliders and share their information with the class.
- Have airplane races to see which student's plane has the longest flight.

Closure

In their magnetism and electricity journals, have students make a design for a glider of their own.

The Big Why

The ruler has a negative charge because it was rubbed with a wool cloth. The aluminum foil airplane is neutral. When the ruler was moved close enough to the airplane, the airplane was attracted to the ruler. After a short while, the airplane became balanced with the ruler. Since like charges repel, the airplane dropped away and began to fly.

Electrostatic Airplane *(cont.)*

Answer the following questions.

1. Why was the ruler attracted to the airplane?

____

____

2. What caused the airplane and the ruler to repel each other?

____

____

3. How long did your airplane stay in the air?

____

____

4. What factors affected the flight of your airplane?

____

____

5. Draw a picture of your airplane flying away from the ruler.

Telltale Potato

Question

How can you determine which end of a battery is negative and which end is positive?

Setting the Stage

- Have students define the terms *anode, cathode*, and *battery*.
- Have students brainstorm all of the possible uses batteries have.
- Create a master list of all the students' ideas.

Materials Needed for Each Group

- a raw potato
- a battery
- two pieces of insulated wire 12" (30 cm) long, with ends exposed
- clear tape
- a knife
- data-capture sheet (page 36), one per student

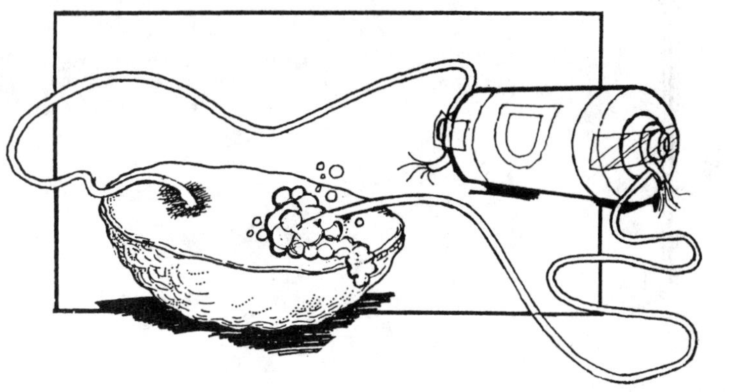

Procedure *(Student Instructions)*

1. Slice the potato in half so there is one flat side.
2. Connect one wire to each of the battery terminals using the tape.
3. Stick the other ends of the wires into the flat side of the potato, about 1" (2.5 cm) apart.
4. Observe what is happening and record on your data-capture sheet.

Extensions

- Have students repeat the experience using a variety of fruits and vegetables.
- Have students repeat the experience placing the wires at different distances apart.

Closure

In their magnetism and electricity journals, have students write down why they think their potato did what it did. Then have them write down the correct response.

The Big Why

A green color appears where the positive wire enters the potato. This is because the negative ions in the potato are neutralized at the positive terminal of the battery. Bubbles will appear where the negative wire enters the potato. The bubbles are caused by the forming of hydrogen gas at the negative terminal of the battery.

Telltale Potato *(cont.)*

Draw what you saw happening to your potato.

Conductor or Insulator

Question

How can you tell conductors from insulators?

Setting the Stage

- Have students define the terms *conductor* and *insulator.*
- Discuss with students the importance of conductors and insulators.
- Have students do research on the history of electricity.
- Refer to diagram on page 69, for the design of a simple circuit.

Materials Needed for Each Group

- three pieces of insulated wire 8" (20 cm) long, with ends exposed
- masking tape
- flashlight bulb
- bulb holder (optional)
- size "D" battery
- a variety of objects (e.g., coin, fork, nail, pencil, toothbrush, wooden ruler)
- data-capture sheet (page 38), one per student

Procedure *(Student Instructions)*

1. Tape one wire to the positive side of the battery.
2. If you have a bulb holder, attach the other end of the wire to one of the bulb holder terminals. Otherwise, wrap the wire around the metal base of the flashlight bulb.
3. Tape the second wire to the negative side of the battery and let the other end of the wire remain unconnected.
4. Connect the third wire either to the remaining terminal of the bulb holder or tape it to the solder drop at the base of the flashlight bulb, separate from the other wire. Leave the other end of the wire unconnected.
5. It is important to see if the battery and light are working before testing your objects. So, take both unattached wires and touch them together to see if the light goes on. If not, check your connections and try again. If it still does not work, check the battery or light. One or both may be faulty.
6. Once you know your equipment is working, you can begin checking your objects one at a time, to see if they are conductors or insulators. Take the bare wires and place them at each end of the object. If the bulb lights up, you have a conductor. If the bulb does not light up, you have an insulator.
7. Record your observations on the data-capture sheet provided.

Extensions

- Have students give oral reports on their electricity research.
- Discuss with students that conductors and insulators can be made from a variety of materials.

Closure

In their magnetism and electricity journals, have students make a list of other objects that can be conductors or insulators and label each object accordingly.

The Big Why

Materials that allow electric current to flow are called conductors. Electric current flows because of the number of free flowing electrons. Materials that do not allow electric current to flow are called insulators.

Conductor or Insulator *(cont.)*

Fill in the information.

Type of Object	Conductor?	Insulator?

The Oh, So Simple Circuit

Question

What happens when a circuit is completed?

Setting the Stage

- Define for students the term *electric current.*
- Discuss with students that in order for electricity to do meaningful work, it must flow through a circuit.
- Have several examples of circuits (simple, series, and parallel) on display for students.

Materials Needed for Each Group

- two "D" cell batteries
- flashlight bulb and holder
- two pieces of insulated wire 12" (30 cm) long, with ends exposed
- pen or pencil
- masking or transparent tape
- data-capture sheet (page 41), one per student

Procedure *(Student Instructions)*

1. Attach an exposed end from each wire to the two ends on the bulb holder. Note: If you just have the bulb, hold one wire to the metal side of the bulb base and the other wire to the bottom of the bulb base.
2. Taking the other ends of the wires, touch the bottom and top of the battery.
3. Record your findings on the data-capture sheet.
4. Place the second battery into your circuit. Note: Make sure the positive end of the second battery is touching the negative end of the first battery. Repeat step 3.
5. Now place both positive ends of the batteries together and complete your circuit. Repeat step 3.
6. Place both negative ends of the batteries together and complete your circuit. Repeat step 3.

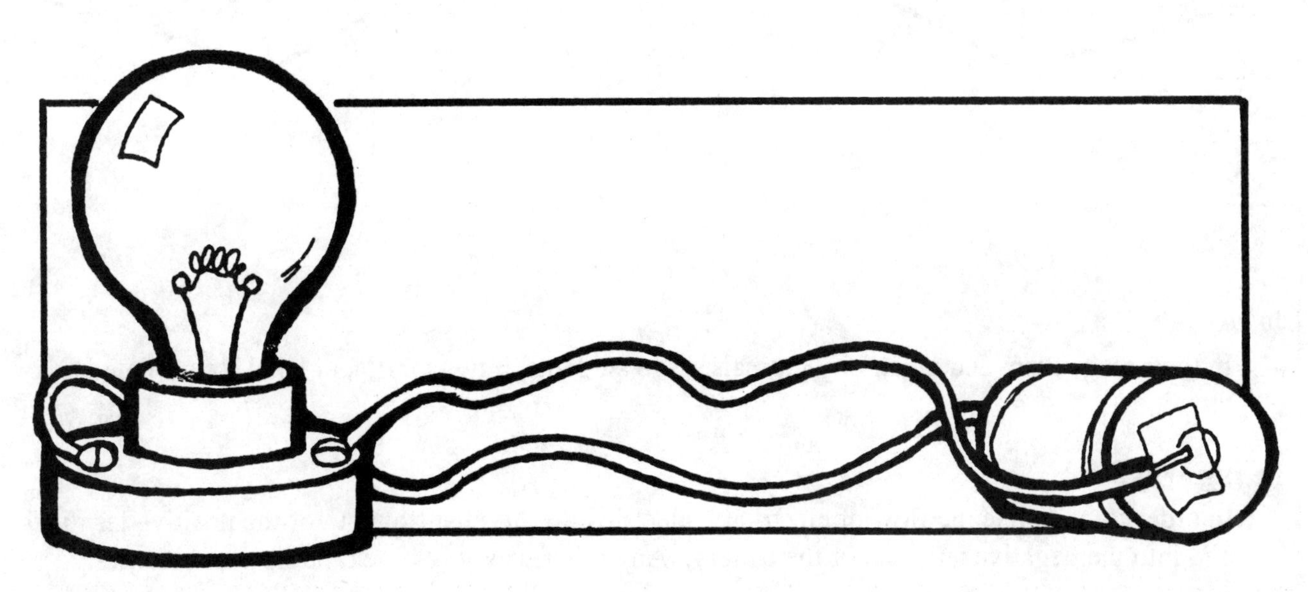

The Oh, So Simple Circuit *(cont.)*

Extensions

- Have students devise as many ways as possible to complete the circuit with the materials they have.
- Using additional materials, see if students can create a series circuit and a parallel circuit.

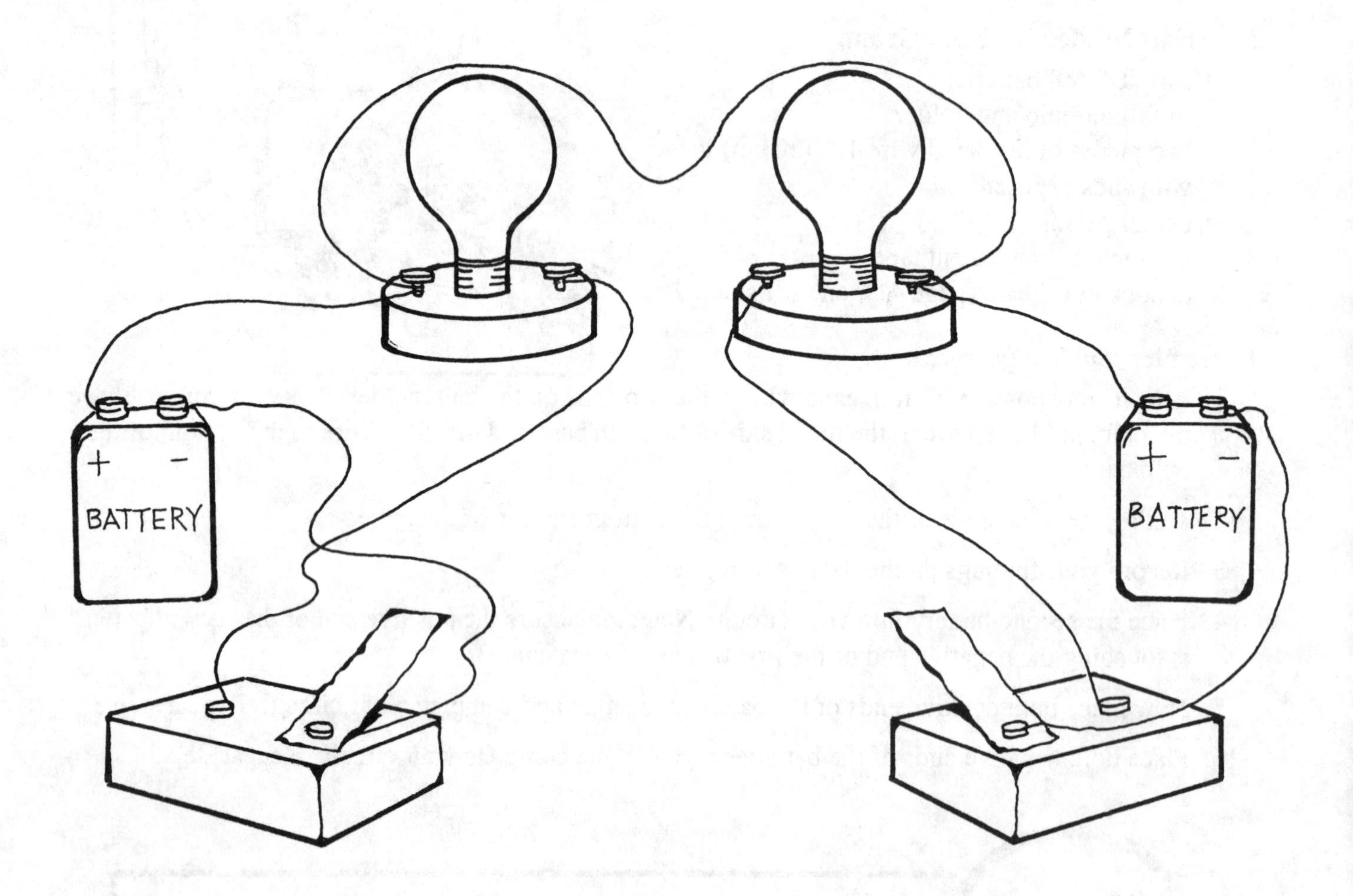

Closure

In their magnetism and electricity journals, have students draw, color, and label a completed circuit.

The Big Why

In order to complete the flow of electricity, electric current must flow out of the positive terminal and into the negative terminal of the battery. Any other way will block the flow of electricity.

The Oh, So Simple Circuit *(cont.)*

Fill in the chart and answer the questions.

	Did the Circuit Work?	Yes	No
1	circuit with one battery		
2	circuit with two batteries, negative attached to positive		
3	circuit with two batteries, negative attached to negative		
4	circuit with two batteries, positive attached to positive		
5	the circuit that you made		

1. Why did the bulb light for circuits one and two above?

___

___

2. Why did the bulb not light for circuits three and four above?

___

___

3. Why or why not did the bulb light for circuit five above?

___

___

Resisting The Electron Flow

Question

How can you control the electron flow in a circuit?

Setting the Stage

- Review with students the concept of conductors and insulators.
- Have students define the term *rheostat.*
- Have students make a list of how they have seen rheostats used. Then have them share their lists to make a master list for the class.

Materials Needed for Each Group

- several pieces of graphite from a mechanical pencil
- three pieces of insulated wire 12" (30 cm) long, with ends exposed
- flashlight bulb
- bulb holder (optional)
- battery
- alligator clips (optional)
- clear tape
- data-capture sheet (page 43), one per student

Procedure *(Student Instructions)*

1. Using the alligator clip or a piece of tape, attach one wire to one side of the graphite.
2. Attach the other end of the wire to one terminal on the bulb holder or wrap it around the metal base of the flashlight bulb.
3. Attach a second wire to the other terminal on the bulb holder or tape it to the solder drop at the bottom of the base of the flashlight bulb, separate from the first wire.
4. Attach the other end of the wire to one terminal of the battery.
5. Attach the third wire to the other battery terminal with the other end of the wire exposed or attached to the other alligator clip.
6. Slide the free end of the third wire along the graphite and observe what happens. Record your observations on your data-capture sheet.

Extensions

Have students repeat the experience by trying other materials. Then have them share their results with the class.

Closure

In their magnetism and electricity journals, have students design a house with all the lights having dimmer switches (rheostats).

The Big Why

The graphite from the pencil does not conduct electricity very well. It is called a semi-conductor. The greater the distance apart the wires are on the graphite, the greater the resistance to electron flow. This causes the light to change from bright to dim and back.

Resisting The Electron Flow *(cont.)*

Answer the following questions.

1. What causes the light to go from bright to dim?

____

____

2. Why do you think this happens? Explain.

____

____

3. Do you think you can use different materials to get the same result? Why?

____

____

4. Can using a dimmer switch at home help to conserve electricity? Explain.

____

____

5. Draw a picture of your rheostat.

Just the Facts

Hans Christian Oersted was doing an experiment for his students with a battery and a piece of wire coil. A compass was sitting next to the wire coil. When the battery power was turned on, a current zipped through the wire. The north pointing compass needle suddenly pointed to the wire. The current flowing through the wire had created a temporary magnet. Oersted had created and discovered the *electromagnet.*

Electricity, therefore, can be used to make strong electromagnets. These electromagnets can be switched on and off as easily as a lightbulb is switched on and off. Looping the wire makes the magnetic field even stronger. And inserting an iron rod down the center of the wire coil creates an even more powerful electromagnet.

Facts about electromagnets.

1. Giant electromagnetic cranes are used to lift tons of iron and steel in scrap metal yards.
2. Without electromagnetic devices there would be no televisions, stereos, computers, or hair dryers. You may add to this list uncountable numbers of other electromagnetic inventions such as microwave ovens.
3. Electromagnetic waves are responsible for all radios and televisions. A moving magnetic field makes up part of a radio wave.
4. Light, which gives us our vision, is also an electromagnetic wave.
5. The strength of an electromagnet may be increased by adding more electrical current to it.
6. The strength of an electromagnet may be increased by winding more coils of wire around its core.

History Repeated

Question

Does electric current produce a magnetic field?

Setting the Stage

- Introduce the experiments done by Hans Christian Orsted in the 1820's.
- Discuss with students the relationship between electricity and magnetism.

Materials Needed for Each Group

- "D" battery
- 1' (30 cm) of insulated copper wire, with the ends exposed
- transparent tape
- compass
- data-capture sheet (page 46), one per student

Procedure *(Student Instructions)*

1. Tape one end of the exposed copper wire to the positive end of the battery.
2. Wrap the wire once around the compass.
3. Take the other end of the exposed copper wire and touch it to the negative end of the battery. Observe what happens to the compass and record on your data-capture sheet.
4. Repeat the experience, this time running the wire under the compass. Observe what happens and record on your data-capture sheet.

Extensions

- Have students repeat the experience, this time reversing the battery connections.
- Have students research some of the early pioneers of electromagnetism and then report to the class.

Closure

In their magnetism and electricity journals, have each student write a paragraph about the discovery of electromagnetism. Each student should pretend to be the one who discovered it.

The Big Why

This experience proves there is a magnetic field found within and around an electric current.

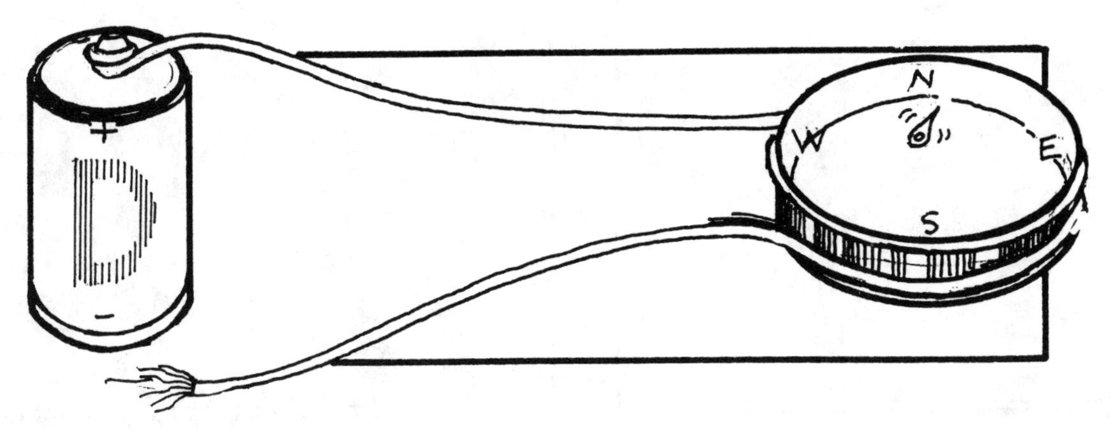

History Repeated *(cont.)*

Answer the following questions.

1. What did you observe when both ends of the copper wire were connected to the battery?

2. Why do you think the compass needle moved?

3. Do you think a stronger battery would have a greater effect on the compass needle? Explain.

4. Do you think a longer piece of wire would have a greater effect on the compass needle? Explain.

5. Draw pictures of what happened when the wire was wrapped around the compass and when the wire was under the compass.

Nail Attractor

Question

What are electromagnets and how are they used?

Setting the Stage

- Discuss with students what electromagnetism is and how it works.
- Display for the class either working models of electromagnets or pictures of them.
- Have students brainstorm a list of possible uses for electromagnets.

Materials Needed for Each Group

- one large iron nail
- one "C" or "D" celled battery
- masking or transparent tape
- 3' (90 cm) of insulated wire with the ends exposed
- a variety of magnetic and nonmagnetic objects (paper clips, coins, pins, keys, etc.)
- pen or pencil
- data-capture sheet (page 48), one per student

Procedure *(Student Instructions)*

1. While holding the nail, wrap the middle 12" (30 cm) of wire tightly, from just below the head to just above the point of the nail.
2. Take the two exposed ends of the wire and attach them to the negative and positive ends of the battery, using either the masking or transparent tape.
3. Once everything has been connected, touch the nail to all the objects that have been collected and see which objects the nail can attract.
4. Record the results on the data-capture sheet.
5. Disconnect the wires from the battery and see if the nail can still pick up any of the objects and for how long.
6. Record the results on the data-capture sheet.

Extensions

- Have students wrap more of the wire around the nail to see if the magnet gets stronger. Add a second battery to the circuit.
- Have students unwrap some of the wire from the nail and retest the metal objects. What is the fewest wraps you can have to still pick up a paper clip?
- Let students create their own electromagnet with materials you give them.

Closure

In their magnetism and electricity journals, have students explain the difference between permanent magnets and electromagnets.

The Big Why

Passing an electric current through certain types of metal objects creates a magnetic field. However, the magnetic field only lasts as long as the electric current is present.

Nail Attractor *(cont.)*

Fill in the chart and answer the following questions with the information gathered from the experience.

Did the Electromagnet pick up the object?	Yes	No	How many?
Did the Electromagnet pick up the _______?			
Did the Electromagnet pick up the _______?			
Did the Electromagnet pick up the _______?			
Did the Electromagnet pick up the _______?			
Did the Electromagnet pick up the _______?			
Did the Electromagnet pick up the _______?			
Did the Electromagnet pick up the _______?			

Did the electromagnet pick up any objects when the wire was disconnected from the battery? If so, for how long?

____

____

____

Electromagnetic Strength

Question

Why would you want to strengthen an electromagnet?

Setting the Stage

- Review with students what an electromagnet is and what it does.
- Discuss with students the use of electromagnets in our society.
- Have students go home and make a list of how electromagnets are used in their homes.

Materials Needed for Each Group

- "D" battery
- 2' (60 cm) of thin, insulated copper wire, with the ends exposed
- several paper clips
- large iron nail
- pen or pencil
- craft knife
- transparent tape
- data-capture sheet (page 50), one per student

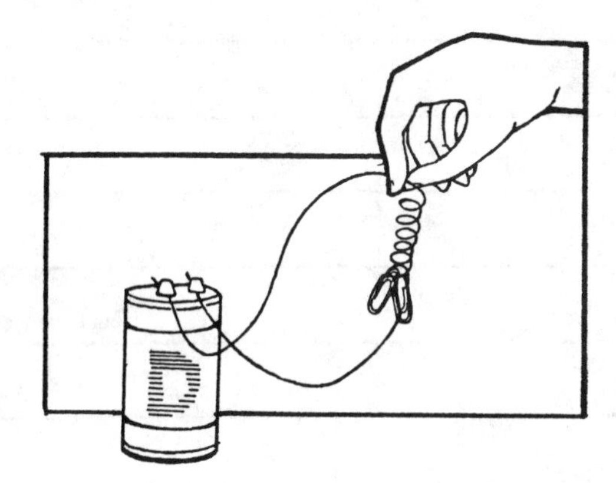

Procedure *(Student Instructions)*

1. Tape one end of the wire to the positive end of the battery and the other end of the wire to the negative end of the battery.
2. Once the wire has been connected to the battery, try to pick up a paper clip from the middle section of the wire. Record your observations on the data-capture sheet provided.
3. This time loosely coil 1' (30 cm) of the wire around a pen or pencil, leaving 6" (15 cm) of wire unwrapped on each side.
4. Remove the pen or pencil from the coil.
5. Using the craft knife scrape off some of the wire insulation from one side of the coil.
6. Connect the wire ends to the battery and try to pick up as many paper clips with the uninsulated part of the coil. Record your observations on the data-capture sheet provided.
7. Finally, take the coil and tightly wrap it around the large iron nail.
8. Connect the wire ends to the battery and try to pick up as many paper clips with the coil-wrapped nail. Record your observations on the data-capture sheet provided.

Extensions

- Have students brainstorm other ways of making a stronger electromagnet and then put those ideas into practice.
- Have a class contest to see who can make the strongest electromagnet.

Closure

In their magnetism and electricity journals, have students come up with an idea for making an electromagnet and then draw it.

The Big Why

As you add more electric current to the electromagnet, it becomes stronger.

Electromagnetic Strength *(cont.)*

Complete the chart and answer the questions.

Type of Test Being Performed	How Many Paper Clips

In which test where you able to pick up the most paper clips? Explain.

____

____

____

Can you think of two ways to make your electromagnet stronger?

____

____

____

Keeping It Current

Question

How else can you create electric current without using a stored source of energy such as a battery?

Setting the Stage

- Ask students if they can think of any way to create an electric current without completing a circuit?
- Make a class list with their answers.
- See if any of the ideas listed are used in today's society.

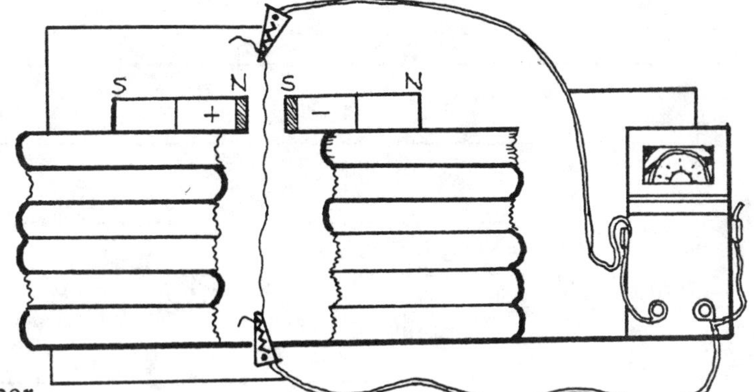

Materials Needed for Each Group

- ammeter or multimeter
- four alligator clips
- 12 text books
- two strong bar magnets
- two 12" (30 cm) lengths of insulated copper wiring, with exposed ends
- 6" (15 cm) length of stripped copper wiring
- data-capture sheet (page 52), one per student

Procedure *(Student Instructions)*

1. Make two piles of books, six high, and place the piles about 3" (7.5 cm) apart from each other.
2. Place one bar magnet on top of each pile of books with their opposite poles facing each other. Each magnet should extend past the books.
3. Attach one alligator clip to each exposed end of insulated wire.
4. Hook two alligator clips to the ammeter or multimeter.
5. Hook the remaining two alligator clips to both ends of the stripped copper wire.
6. Move the copper wire vertically up and down between the two magnets.
7. Observe what is happening and record the information on your data-capture sheet.

Extensions

- Have students repeat the experience, by moving the stripped copper wire at different speeds.
- Have students try materials other than copper wire and have them record their results.

Closure

In their magnetism and electricity journals, have students write down a general statement about what they learned.

The Big Why

Moving the copper wire vertically through the magnetic field induces a small electric current.

Keeping It Current *(cont.)*

Answer the following questions.

1. What did you observe when you moved the copper wire through the magnetic field?

____

____

____

2. Why do you think this happened?

____

____

____

3. What was the reading of the ammeter or multimeter?

____

____

____

4. Do you think you can increase or decrease the reading on the meter? How?

____

____

____

5. List some ways this could be put to practical use.

____

____

____

Buzz

Question

What makes an electromagnetic buzzer buzz?

Setting the Stage

- Discuss with students some of the common places you find electromagnetic buzzers.
- Have students go home and find all the uses of buzzers in the house.
- Make a class list with all of the uses.

Materials Needed for Each Group

- 6-volt battery
- two pieces of wood, one piece 2" x 6" (5 cm x 15 cm) and one piece 1" x 3" (2.5 cm x 7.5 cm)
- large iron nail, 3.5" (8.75 cm) long
- metal nail file with a hole at one end
- thumb tack
- two 4' (1.22 m) lengths of insulated copper wire, with the ends exposed
- glue (wood)
- data-capture sheet (page 54), one per student

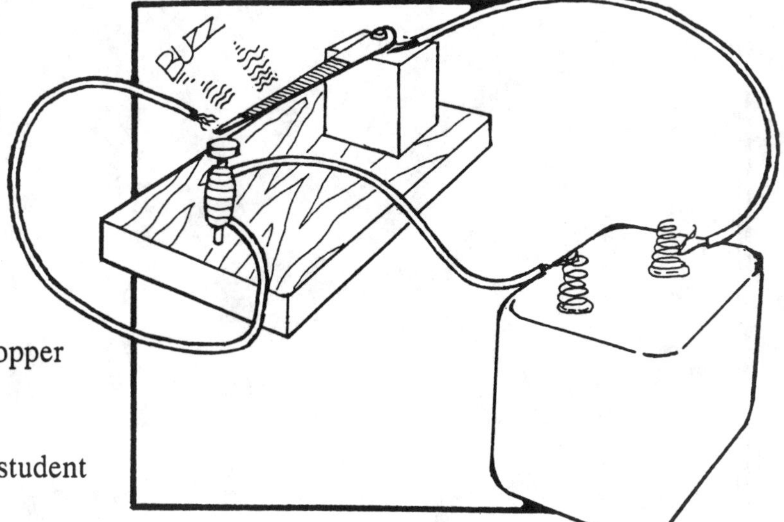

Procedure *(Student Instructions)*

1. Glue both pieces of wood together into an "L" shape and allow time to dry.
2. Hammer the nail into the far side of the "L," about a ¼" (.64 cm) below the top of the short leg "L," making sure the nail file can still touch the top of the nail.
3. Take one length of wire and wrap the middle 2' (60 cm) of wire around the nail.
4. Place the nail file on top of the small piece of wood and use the thumb tack to keep it in place. The end of the nail file should be ¼" above the nail head. If not, adjust the nail.
5. Connect one of the wire ends wrapped around the nail to one terminal of the battery.
6. Take the second wire and attach one end to the nail file and the other end to the second terminal of the battery.
7. With the remaining wire end from the nail, gently touch the end of the nail file. Observe what happens and record on your data-capture sheet.

Extensions

- Discuss with students why the electromagnet buzzes.
- Have students create another design for an electromagnet buzzer and then try to make it.

Closure

In their magnetism and electricity journals, have students describe in their own words why the electromagnet buzzes.

The Big Why

When the bare wire from the electromagnet touches the nail file, the circuit has been completed, creating a magnetic field. This magnetic field attracts the nail file towards the electromagnet, breaking the circuit, whereupon the nail file springs back up to the wire and the process repeats itself. As the nail file springs back-and-forth, it creates a buzzing sound.

Buzz *(cont.)*

Draw a picture of what happens when the circuit from the buzzer is completed.

Simply An Electric Meter

Question

What are some of the uses for an electric meter?

Setting the Stage

- Talk with the class about Oliver B. Shallenberger, the pioneer of the electric meter.
- Have students research some of the uses electric meters have in today's society.
- Then have them share their information with the class.

Materials Needed for Each Group

- cardboard tube (from paper towel or toilet paper)
- flashlight bulb
- bulb holder (optional)
- small magnetic compass
- "D" battery
- transparent tape
- two 2.5' (75 cm) lengths of insulated copper wire, with the ends exposed
- data-capture sheet (page 56), one per student

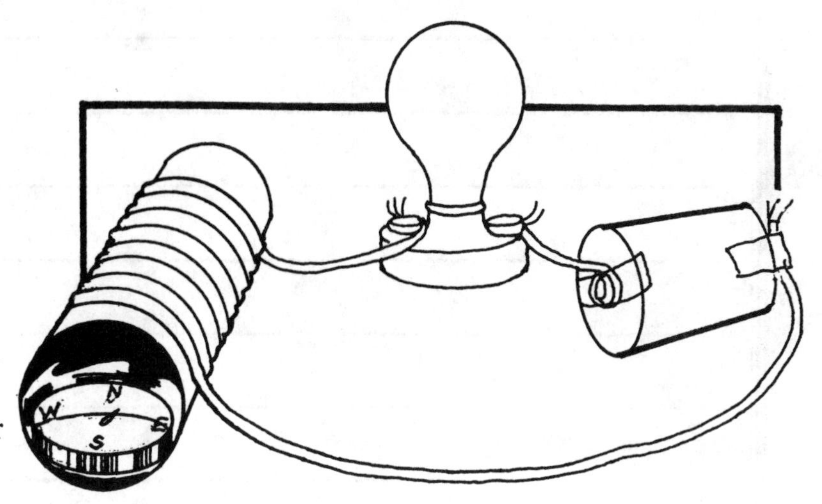

Procedure *(Student Instructions)*

1. Wrap one length of copper wire around the cardboard tube. Make sure to leave about 8" (20 cm) of wire free on each end.
2. Tape one of the free ends of wire to one side of the battery.
3. Connect the other free end of wire to one terminal of the bulb holder or wrap it around the base of the bulb.
4. Place the compass inside the cardboard tube, where it is still visible.
5. Tape one end of the second piece of wire to the other side of the battery.
6. Attach the remaining end of wire to the other bulb holder terminal or wrap it around the base of the bulb, making sure to not touch the other wire.
7. When all of the connections have been made, observe what is happening to the compass and record the information on your data-capture sheet.

Extensions

- Have students repeat the experience, first wrapping less wire around the tube and then wrapping more around the tube.
- Bring in an electric bill for the class and show them how the bill is determined.

Closure

In their magnetism and electricity journals, have students draw a picture of their electric meter.

The Big Why

The electric current in the wire creates a magnetic field, causing the compass needle to move.

Simply An Electric Meter *(cont.)*

Write a paragraph describing what is happening to the compass when all the connections have been made.

Mr. Morse, I Presume

Question

How are electromagnets used in communication?

Setting the Stage

- Discuss with the class early methods of communication.
- Have students list ways communication has changed because of electricity. Then make a class list using the students' ideas.
- Hand out copies of Morse code to everyone in the class.

Materials Needed for Each Group

- one 6 volt battery or several "D" cell batteries
- one paper clip
- three thumb tacks
- flashlight bulb and holder
- two wood blocks—one 3" (7.5 cm) long and one 4" (10 cm) long
- white glue
- 3" (7.5 cm) iron nail
- a 1" x 3" (2.5 x 7.5 cm) strip of tin
- a 1.5" x 3" (3.75 cm x 5 cm) piece of wood
- two 1' (30 cm) lengths of insulated wire and one 3' (90 cm) length of wire, all with the ends exposed 1" (2.5 cm)
- data-capture sheet (page 59), one per student

Procedure *(Student Instructions)*

1. Glue the 3" (7.5 cm) block on top of the 4" (10 cm) block in the shape of an "L." Allow 24 hours to dry.
2. Lay the "L" on a table so the bottom of the "L" sticks up in the air. Hammer the 3" (7.5 cm) iron nail into the part of the "L" that is on the table surface, 2" (5 cm) from the part sticking in the air.
3. Paper clip switch: Take the wood, paper clip, and two thumb tacks. Bend out the paper clip so it looks like an "S." Tack down the large side of the "S" to one end of the wood. Put the second tack on the other side of the wood and bend the small side of the "S" so it is just above the second tack.

Mr. Morse, I Presume *(cont.)*

4. Wrap the middle 12" (30 cm) of the 3' (90 cm) wire tightly around the nail, from top to bottom.
5. Run one of the wire ends from the nail to the terminal of the bulb holder and the other wire end under the tack on one side of the paper clip switch.
6. Using a 1' (30 cm) piece of wire, attach one end to the bulb holder and the other end to one of the battery terminals.
7. Using the other 1' (30 cm) piece of wire, attach one end to the other battery terminal and the other end under the tack on the other side of the paper clip switch.
8. Using the remaining thumb tack, fasten the strip of tin to the top of the "L" that is sticking in the air.
9. Push the paper clip down and make contact with the tack to work the telegraph.

Extensions

- Once the telegraph is complete, have students send each other messages using Morse code.
- Have students connect two telegraphs and send messages from the other side of the room or another room.
- Have students use a buzzer instead of a flashlight bulb.
- Have students research about early telegraphs.

Closure

In their magnetism and electricity journals, have students make up their own code for their telegraphs.

The Big Why

When the paper clip switch touches the second thumb tack, the circuit is completed, activating the electromagnet. The bulb lights and the metal strip is pulled towards the nail. When the paper clip is released, the circuit is broken and the electric current is stopped.

Mr. Morse, I Presume *(cont.)*

Translate the Morse code messages using the chart below.

1. ____

2. The inventor of Morse code was

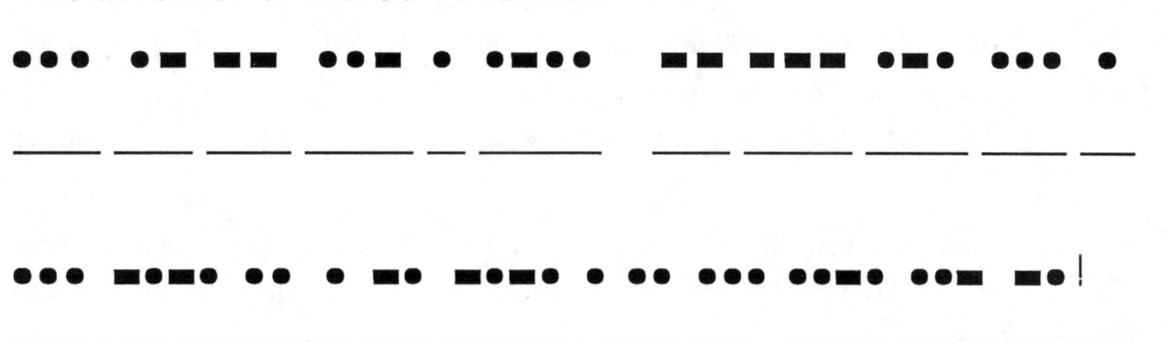

3. ____ !

4. Write your own message.

Morse Code

a	·–	s	···
b	–···	t	–
c	–·–·	u	··–
d	–··	v	···–
e	·	w	·––
f	··–·	x	–··–
g	––·	y	–·––
h	····	z	––··
i	··	1	·––––
j	·–––	2	··–––
k	–·–	3	···––
l	·–··	4	····–
m	––	5	·····
n	–·	6	–····
o	–––	7	––···
p	·––·	8	–––··
q	––·–	9	––––·
r	·–·	0	–––––

Language Arts

Reading, writing, listening, and speaking experiences blend easily with the teaching and reinforcement of science concepts. Science can be a focal point as you guide your students through poems and stories, stimulating writing assignments, and dramatic oral presentations. If carefully chosen, language arts material can serve as a springboard to a magnetism and electricity lesson, the lesson itself, or an entertaining review.

There is a wealth of good literature to help you connect your curriculum. Some excellent choices are suggested in the Bibliography (pages 95-96).

Science Concept: *Electricity and magnetism have a variety of uses.*

- Have students imagine they are a superheroes who have magnetic powers. Have them write a short story about their adventures as a magnetic superhero. (A villain named "Magneto" appeared in Marvel Comics.)

- Have a discussion with your students about how electricity is delivered to their homes from the generating station. Then have students draw a picture of what this might look like.

- Have student form into groups. Have each group write about what happens at each stage of electrical power delivery. Have each group do an oral presentation of their stage.

Social Studies

Magnetism and electricity have played a significant role in history. Cultures have been built around their use, and people have devoted their lives to working with them to make conditions in the world better in some way.

As you guide your students through lessons in history, geography, cultural awareness, or other areas of social studies, keep in mind the role magnetism and electricity have played. You will find it easy to incorporate the teaching and reinforcement of science concepts in your lessons.

Science Concept: *Electricity plays an important role in our daily lives.*

- Have students brainstorm a list of things in their lives that use electricity.
- Have students cut out magazine pictures of things that need electricity to power them. These can be made into a poster.
- Ask students for their ideas about what would happen if they did not have a refrigerator, stove, or air conditioner in their homes.
- Have students research who some of the great inventors of magnetism and electricity were. Which countries did they come from? Indicate these countries with markers on a world map.
- Ask students how people live in different countries where there is very little electricity.
- Have students research which countries in the world use the most energy. Where does their country rank on this list?
- Telephones are powered by electricity. Have students find out all they can about telephones. Who invented the telephone? What would their lives be like without a telephone?

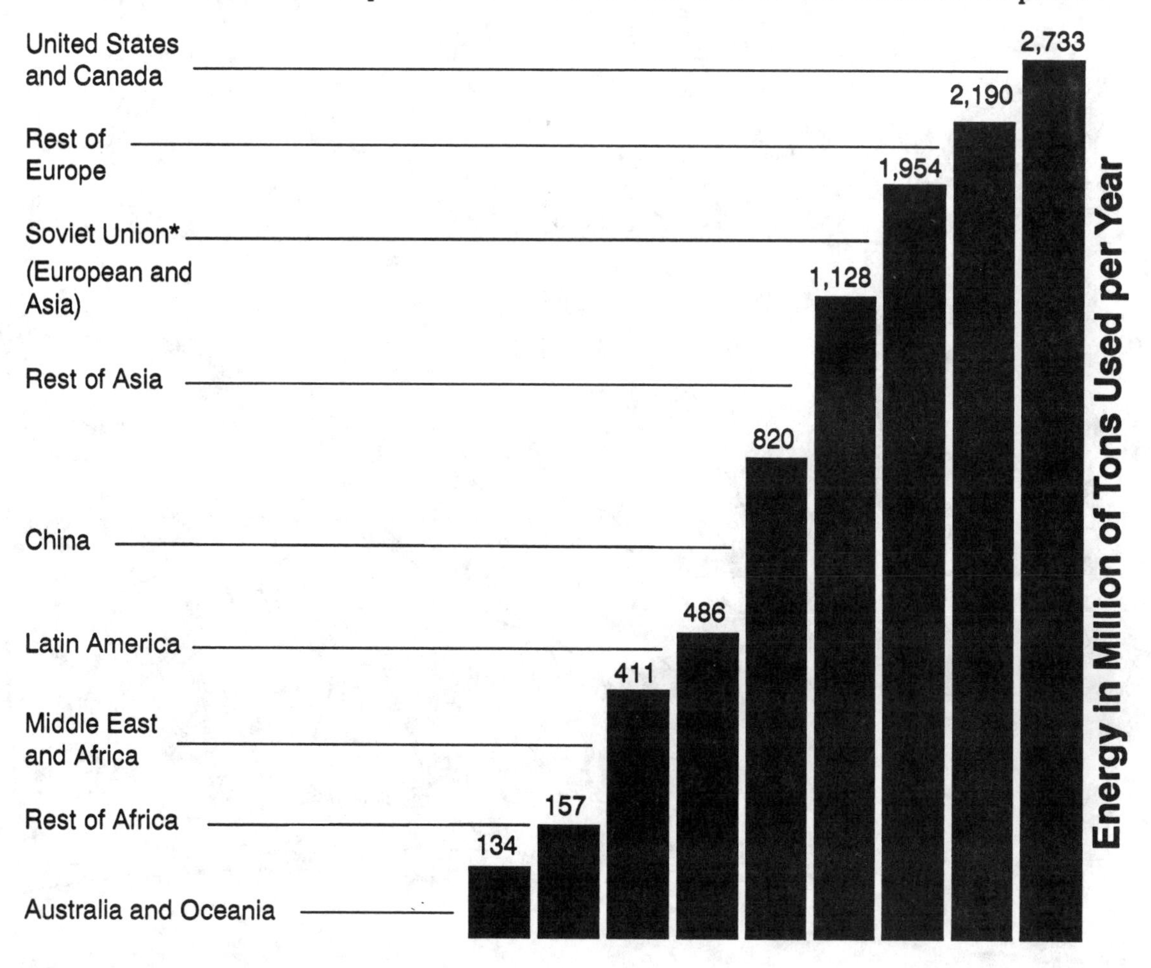

*Figures are for 1988 prior to the breakup of the Soviet Union

Physical Education

What can be more fun for primary students then pretending they are an electrical current, subject to the forces that make electrical currents possible. Here is an opportunity to let your students develop their knowledge of magnetism and electricity in a physical way.

Science Concept: *Electric current moves in a continuous stream.*

Divide the class into two groups. These groups need to form straight lines which will represent electrical wires. Give a ball to the student at the front of each line. The ball will represent the current. The object is to pass the ball back through their legs. When the last persons in line get the ball, they come up to the front of the line and start the ball moving again. The relay continues until the person who started the relay is back at the front of the line.

Physical Education *(cont.)*

Science Concept: *Magnetic force can make other objects into magnets.*

Have students play a game of magnetic tag. One person is designated the magnet, while everyone else needs to keep away. As the magnet tags other people, they also become magnetized and have to join the magnet by holding hands. Now the two magnets act as one as they try to tag other people. As more and more people are tagged, they also have to hold on to the group of magnets. When the group becomes too large, they are allowed to separate into more than one group (e.g., the sixth person tagged may become the lone magnet.) This game is best played in a large open area with marked boundaries.

Math

The study of magnetism and electricity requires the use of math-oriented skills. The ability to measure, compare, and graph are just a few of the skills that can bring mathematics into your magnetism and electricity lessons.

- Teach or review the use of measuring tools (such as rulers with centimeters and inches to measure length.)
- Have students practice reading and making charts and graphs.
- Provide opportunities for students to record data on a variety of graphs and charts. Teach the skills necessary for success.
- Encourage students to devise their own ways to show the data they have gathered.
- On an appropriate level, teach how to average test results.
- Challenge students to find mathematical connections as they study magnetism and electricity.

Science Concept: *Investigations can be repeated to verify results.*

Materials Needed

- several magnets (bar, horseshoe, etc.)
- paper clip
- graph paper
- ruler (standard or metric)

Note to the teacher: Have students measure the strength of a magnet.

Procedure *(Student Instructions)*

1. Place a magnet on a piece of graph paper.
2. Next, place a paper clip on the opposite side of the paper.
3. Then you should move the magnet one unit closer to the paper clip at a time and record the distance when the paper clip is affected by the magnet.
4. Finally, repeat this experience with several different magnets and record the results.

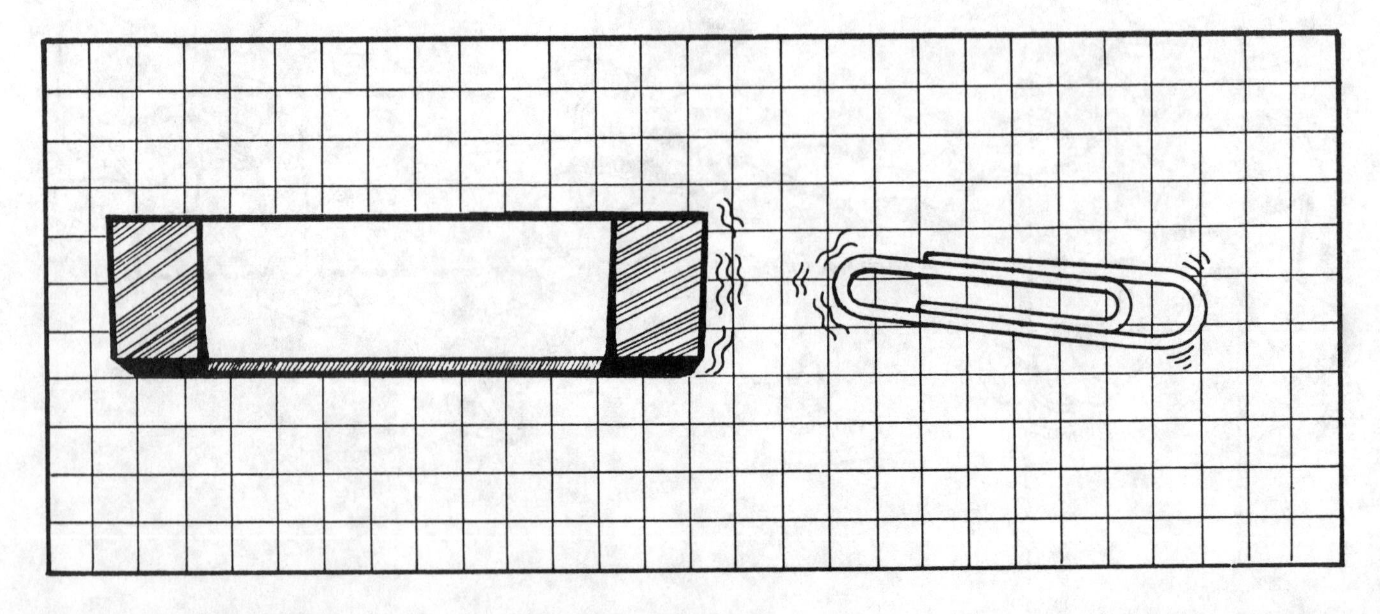

Art

Art projects using magnets and electricity can be done easily at school and home. Your students will enjoy all of the possibilities.

Science Concept: *A magnetic force passes into a magnetically attracted object, making it magnetic as well.*

Materials Needed

- several strong magnets
- various small objects that are attracted to magnets (e.g., washers, paper clips, nails, small pieces of metal, etc.)

Have students make a "magnetic" sculpture from the materials available.

Music

Electricity is a very important part of the music industry. Many musical instruments require electricity for them to work. Much of today's music is also orchestrated with the use of electronic instruments such as synthesizers or mixers.

Science Concept: ***Electricity plays an important role in the use of many musical instruments.***

- Have students collect as many pictures as they can of instruments that require electricity and make a collage of them.
- Bring into class recordings of instruments that require electricity, such as the electric guitar, electric keyboard, electric bass, etc.
- Show students videos of bands that use electronic instruments.
- Invite a person from a music store to come into class and demonstrate a variety of electrical instruments.

Observe

This activity will demonstrate whether a magnetic force can be stopped. As you do the activity, observe which materials can stop the magnetic force.

Materials Needed at This Station

- bar magnet
- large paper clip
- several sheets of paper
- several pieces of cloth
- plastic bags
- cardboard box
- wire screen
- aluminum foil
- data-capture sheet (page 68), one per student

Procedure *(Student Instructions)*

1. Wrap your bar magnet in a piece of cloth and then try to pick up the paper clip. Observe whether the magnet picked up the paper clip and record your observations on your data-capture sheet.
2. Repeat step 1, but this time wrap additional pieces of cloth around the magnet. Observe whether the magnet picked up the paper clip and record your observations on your data-capture sheet.
3. Repeat steps 1-2 using the remainder of the materials on the table at this station. Make sure to record all of your observations on your data-capture sheet.
4. Put your finished data-capture sheet in the collection pocket on the side of the table at this station.

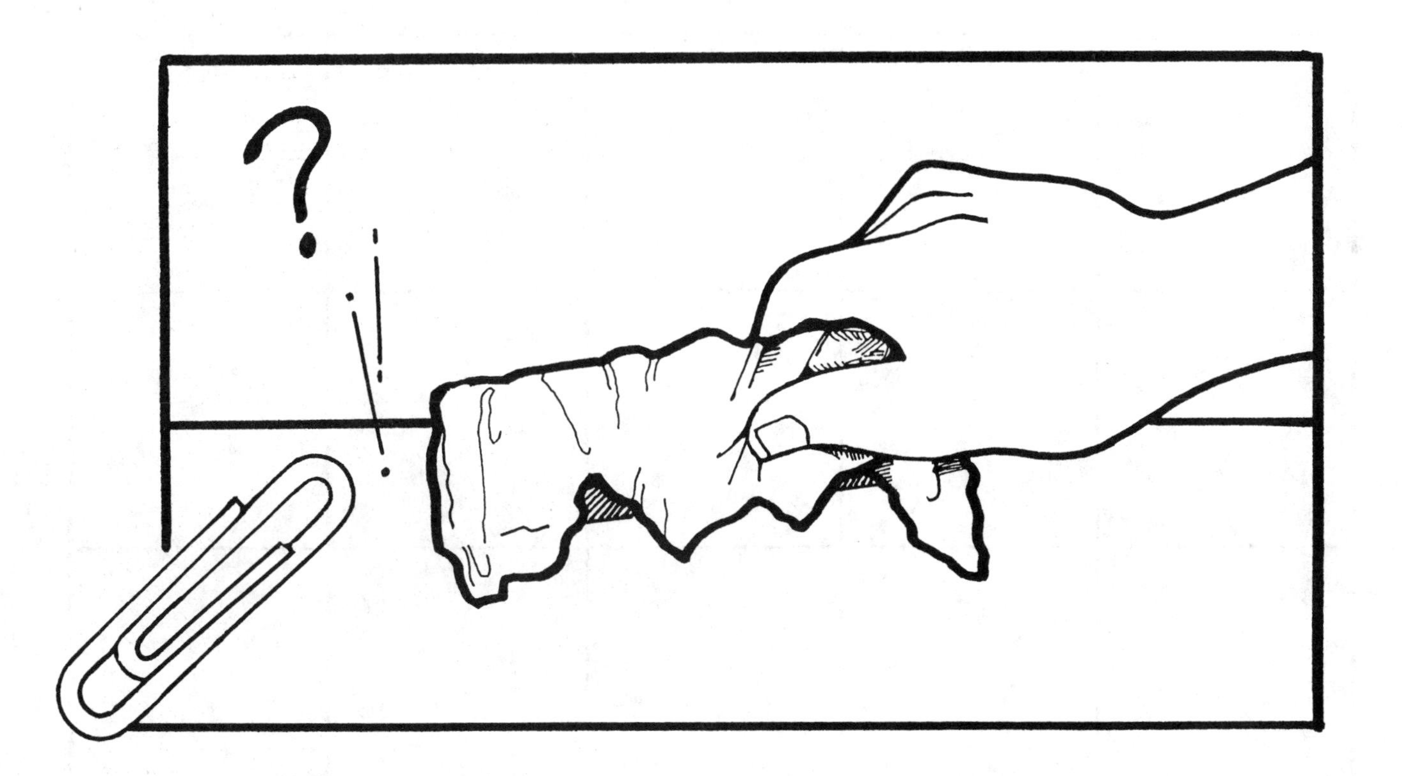

Observe *(cont.)*

Before beginning your investigation, write your group members' names on the lines below.

____________ Project Leader ____________ Stenographer

____________ Physicist ____________ Transcriber

Complete the chart.

Paper	Plastic Bag	Aluminum Foil	Box	Screen

Communicate

At this station you will find ten items (aluminum foil, eraser, copper wire, nail, paper, paper clip, pencil, string, cup of water, piece of wood), that will or will not conduct electricity in a simple circuit. Test each item and record the results.

Materials Needed at This Station

- "D" battery
- three pieces of copper wire 6" (15 cm) long
- flashlight bulb and holder
- two alligator clips
- test items (aluminum foil, eraser, copper wire, nail, paper, paper clip, pencil, string, cup of water, piece of wood)
- data-capture sheet (page 70), one per student

Procedure *(Student Instructions)*

1. One at a time attach the alligator clips to each end of the ten items. If the lightbulb lights up, you have created a simple circuit.
2. Observe what happens during each test and record your observations on your data-capture sheet.
3. Put your finished data-capture sheet in the collection pocket on the side of the table at this station.

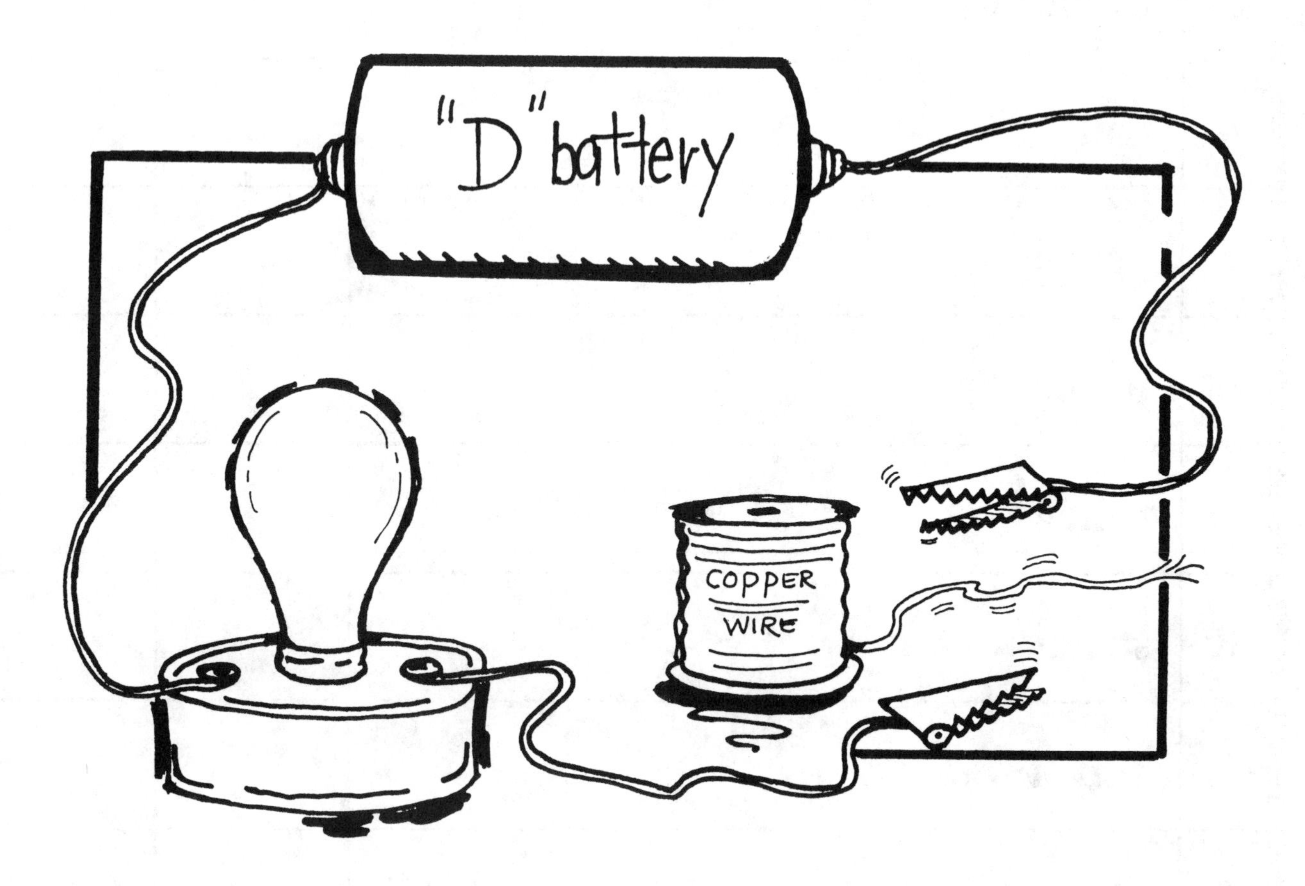

Communicate *(cont.)*

Before beginning your investigation, write your group members' names on the lines below.

_____________ Project Leader _____________ Stenographer

_____________ Physicist _____________ Transcriber

Complete the chart.

	Was the circuit completed?	Yes	No
1	aluminum foil		
2	eraser		
3	copper wire		
4	nail		
5	paper		
6	paper clip		
7	pencil		
8	string		
9	cup of water		
10	piece of wood		

Compare

At this station you will be testing the strength of different magnets.

Materials Needed at This Station

- several different magnets
- several paper clips
- data-capture sheet (page 72), one per student

Procedure *(Student Instructions)*

1. Work together with your group to come up with as many different ways as possible to test the strength of magnets with the materials you have available.
2. Once you have come up with some possibilities, begin testing them. Record your observations on your data-capture sheet.
3. After you have completed all of your tests, choose which test was the best and mark it on your chart.

Procedure *(Teacher Instructions)*

1. When students have finished all of their tests, show them one more possibility. Bend a paper clip into a hook. Attach it to a magnet and add paper clips to the hook until it will not hold anymore.
2. Ask your students if this test is better than the ones they tried.
3. Have students add this test to their charts.
4. Have them put their finished data-capture sheets in the collection pocket on the side of the table at this station.

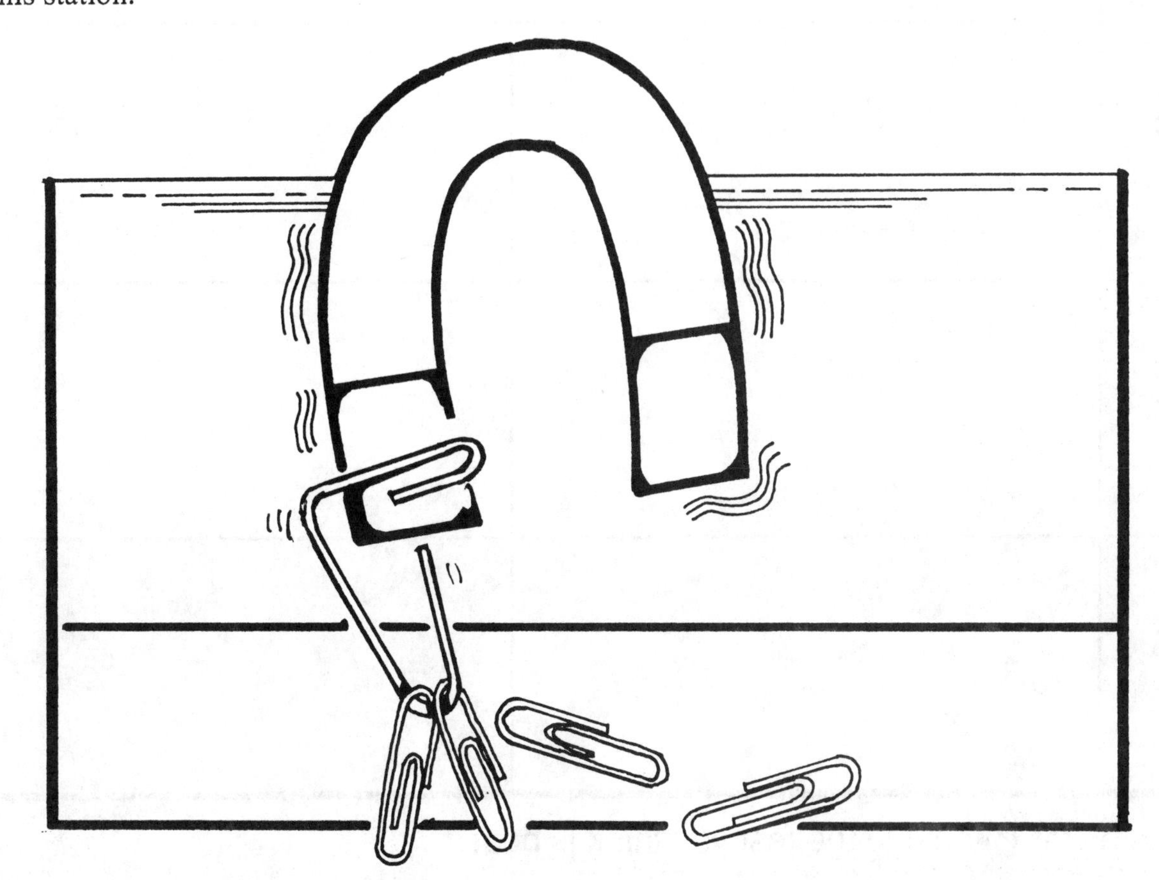

Compare *(cont.)*

Before beginning your investigation, write your group members' names on the lines below.

____________ Project Leader	____________ Stenographer
____________ Physicist	____________ Transcriber

Complete the chart.

	Testing Magnetic Strength	Number of Paper Clips
1		
2		
3		
4		
5		

Place a star (*) next to the test you think is best.

Teacher Note: See page 10 for instructions.

Order

At this station you will be fishing for math problems, with a magnetic fishing pole. Once you get the math problem, solve it and record the answers in numerical order from largest to smallest and then smallest to largest.

Materials Needed at This Station

- fishing pole (pencil, ruler, or stick, string, and a magnet)
- math problems (on cards with washers attached to them)
- a bucket
- data-capture sheet (page 74), one per student

Put your finished data-capture sheet in the collection pocket on the side of the table at this station.

Order *(cont.)*

Before beginning your investigation, write your group members' names on the lines below.

_______________ Project Leader _______________ Stenographer

_______________ Physicist _______________ Transcriber

Complete the chart.

Smallest to Largest

1	2	3	4	5
6	7	8	9	10

Largest to Smallest

1	2	3	4	5
6	7	8	9	10

Put your finished activity paper in the collection pocket on the side of the table at this station.

Categorize

Before beginning your investigation, write your group members' names on the lines below.

_____________ Project Leader   _____________ Stenographer

_____________ Physicist   _____________ Transcriber

At this station you will need to decide which electrical items run on batteries, electricity, or both. Fill in the chart below with this information.

Put your finished activity paper in the collection pocket on the side of the table at this station.

Item	Just Batteries	Just Current Electricity	Both

Teacher Note: See page 10 for instructions.

Relate

Before beginning your investigation, write your group members' names on the lines below.

_______________ Project Leader

_______________ Stenographer

_______________ Physicist

_______________ Transcriber

At this station you will see different pictures of items that run on battery power and items that run on current electricity generated from power stations. Record the items that have similar power sources in the same columns.

Put your finished activity paper in the collection pocket on the side of the table at this station.

Electricity from Power Station	Battery Power

Infer

Before beginning your investigation, write your group members' names on the lines below.

_____________ Project Leader	___________ Stenographer
_____________ Physicist	___________ Transcriber

At this station you will find five drawings of circuits. Your job is to use all of your knowledge of electrical circuits and determine which circuits will light the bulb and which will not.

Put your finished activity paper in the collection pocket on the side of the table at this station.

 Teacher Note: See page 10 for instructions.

Apply

Before beginning your investigation, write your group members' names on the lines below.

______________ Project Leader ______________ Stenographer

______________ Physicist ______________ Transcriber

At this station you will be creating three different working circuits, with the materials provided. After you have made a working circuit, draw a picture in one of the boxes provided.

Materials Needed at This Station

- three flashlight bulbs in holders
- one flashlight bulb not in a holder
- eight 6" (15 cm) pieces of copper wire, with ends exposed
- two "D" batteries
- masking or transparent tape

Circuit #1

Circuit #2

Circuit #3

Put your finished activity paper in the collection pocket on the side of the table at this station.

Science Safety

Discuss the necessity for science safety rules. Reinforce the rules on this page or adapt them to meet the needs of your classroom. You may wish to reproduce the rules for each student, or post them in the classroom.

1. Begin science activities only after all directions have been given.
2. Never put anything in your mouth unless it is required by the science experience.
3. Always wear safety goggles when participating in any lab experience.
4. Dispose of waste and recyclables in proper containers.
5. Follow classroom rules of behavior while participating in science experiences.
6. Review your basic class safety rules every time you conduct a science experience.

You can still have fun and be safe at the same time!

Magnetism and Electricity Journal

Magnetism and Electricity Journals are an effective way to integrate science and language arts. Students are to record their observations, thoughts, and questions about past science experiences in a journal to be kept in the science area. The observations may be recorded in sentences or sketches which keep track of changes both in the science item or in the thoughts and discussions of the students.

Magnetism and Electricity Journal entries can be completed as a team effort or an individual activity. Be sure to model the making and recording of observations several times when introducing the journals to the science area.

Use the student recordings in the Magnetism and Electricity Journals as a focus for class science discussions. You should lead these discussions and guide students with probing questions, but it is usually not necessary for you to give any explanation. Students come to accurate conclusions as a result of classmates' comments and your questioning. Magnetism and Electricity Journals can also become part of the students' portfolios and overall assessment program. Journals are valuable assessment tools for parent and student conferences as well.

How To Make a Magnetism and Electricity Journal

1. Cut two pieces of 8.5" x 11" (22 cm x 28 cm) construction paper to create a cover. Reproduce page 81 and glue it to the front cover of the journal. Allow students to draw magnetism and electricity pictures in the box on the cover.
2. Insert several Magnetism and Electricity Journal pages. (See page 82.)
3. Staple together and cover stapled edge with book tape.

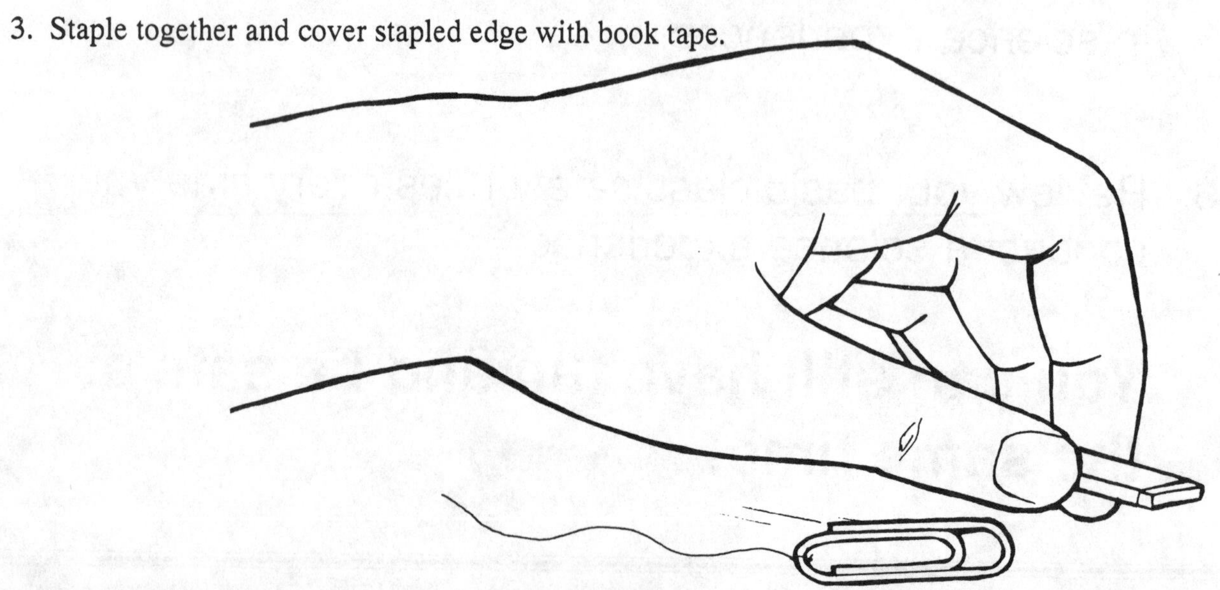

My Magnetism and Electricity Journal

Name ______________________________

Magnetism and Electricity Journal

Illustration

This is what happened: ________________________________

____

____

____

____

This is what I learned: ________________________________

____

____

____

____

My Science Activity

K-W-L Strategy

Answer each question about the topic you have chosen.

Topic: ______________________________

K - What I Already **Know:** ______________________________

____

____

____

____

W - What I **Want to Find Out:** ______________________________

____

____

____

____

L - What I **Learned After Doing the Activity:** ______________________________

____

____

____

____

Investigation Planner *(Option 1)*

Observation

____

____

Question

____

____

Hypothesis

____

____

Procedure

Materials Needed:

____

____

____

____

Step-by-Step Directions: (Number each step!)

____

____

____

____

____

____

____

____

Investigation Planner *(Option 2)*

Science Experience Form

Scientist ______________________________

Title of Activity ______________________________

Observation: What caused us to ask the question?

____

____

____

Question: What do we want to find out?

____

____

____

Hypothesis: What do we think we will find out?

____

____

____

Procedure: How will we find out? *(List step by step.)*

1. ______________________________
2. ______________________________
3. ______________________________
4. ______________________________

Results: What actually happened?

____

____

____

Conclusions: What did we learn?

____

____

Magnetism and Electricity Observation Area

In addition to station-to-station activities, students should be given other opportunities for real-life science experiences. For example, old crank-type telephone magnets or dismantled car generators can provide a vehicle for discovery learning if students are given time and space to observe them,

Set up a magnetism and electricity observation area in your classroom. As children visit this area during open work time, expect to hear stimulating conversations and questions among them. Encourage their curiosity but respect their independence!

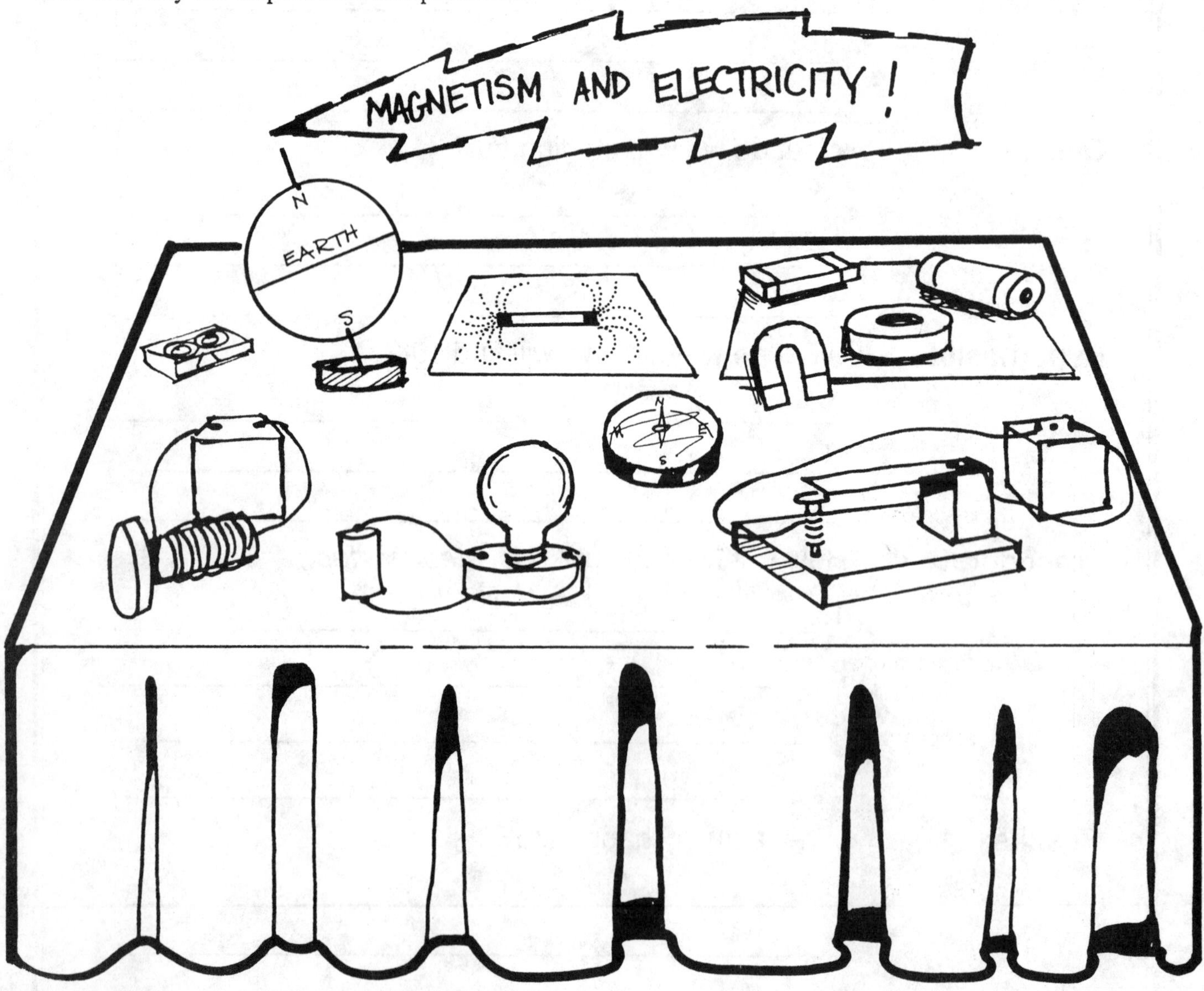

Books with facts pertinent to the subject, item, or process being observed should be provided for students who are ready to research more sophisticated information.

Sometimes it is very stimulating to set up a science experience or add something interesting to the Magnetism and Electricity Observation Area without a comment from you at all! If the experiment or materials in the observation area should not be disturbed, reinforce with students the need to observe without touching or picking up.

Assessment Forms

The following chart can be used by the teacher to rate cooperative-learning groups in a variety of settings.

Science Groups Evaluation Sheet

Room: ____________________ Date: ____________________

Activity: ____

Everyone	Group 1	2	3	4	5	6	7	8	9	10
. . . gets started.										
. . . participates.										
. . . knows jobs.										
. . . solves group problems.										
. . . cooperates.										
. . . keeps noise down.										
. . . encourages others.										

Teacher comment

____

Bragging rights for the group session: ____________________

____

Assessment Forms *(cont.)*

The evaluation form below provides student groups with the opportunity to evaluate the group's overall success.

Cooperative Group Evaluation

Assignment: ______________________________

Date: ________________

Scientists	**Jobs**
________________	________________
________________	________________
________________	________________
________________	________________

As a group, decide which face you should fill in and complete the remaining sentences.

1. We finished our assignment on time, and we did a good job.
2. We encouraged each other, and we cooperated with each other.
3. We did best at ______________________________

   ______________________________.
4. Next time we could improve at ______________________________

   ______________________________.

Assessment Forms *(cont.)*

The following form may be used as part of the assessment process for hands-on science experiences.

Science Anecdotal Record Form

Date: ______________________________

Scientist's Name: ______________________________

Topic: ______________________________

Assessment Situation: ______________________________

Instructional Task: ______________________________

Behavior/Skill Observed: ______________________________

____

____

____

____

____

This behavior/skill is important because ______________________________

____

____

____

____

______________________________.

Creating Science Projects

At the end of each lesson in this book, have students think about questions that were left unanswered or that they would enjoy investigating further. Help them focus their questions into science project investigations. The following example shows how the process that is used throughout the book may be used in the creation of science projects.

Example

Teacher: What did you learn in our science lesson today?

Student: Electromagnets need to have an electric current to work.

Teacher: What other questions about electromagnets and how they work would be really interesting to you?

Student: I want to know if adding more batteries to an electromagnet circuit makes a stronger electromagnet.

Once students decide which question they would like to investigate, have them use the scientific method to do it. A project stemming from the above question may end up looking something like this:

Question

Will adding more batteries to an electromagnet circuit make a stronger electromagnet?

Hypothesis

Adding more batteries to the electromagnet circuit will make a stronger electromagnet.

Materials Needed for Each Individual

- large iron nail
- three "D" batteries
- masking or transparent tape
- 3' (90 cm) of insulated copper wire, with exposed ends
- paper clip
- ruler
- data-capture sheet (page 91)

Procedure *(Student Instructions)*

1. While holding the nail, wrap the middle 12" (30 cm) of wire tightly from just below the head to just above the point of the nail.
2. Take the two exposed ends of the wire and attach them to the positive and negative ends of a battery.
3. Put the paper clip on the ruler at the 1" (2.5 cm) mark.
4. Starting on other side of the ruler, slowly move the electromagnet closer to the paper clip until the paper clip jumps to the electromagnet. Repeat this a second time to verify your results. Then, record your results on your data-capture sheet.
5. Add a second battery to the circuit by taping it in place.
6. Repeat steps 3 and 4.
7. Add a third battery to the circuit by taping it in place.
8. Repeat steps 3 and 4.

Glossary

Alternating Current—electric current that is constantly changing its direction of flow.

Ammeter—a special meter used for measuring electric current.

Anode—an electrode that is positively charged.

Atom—the smallest part of a substance. Inside the atom you will find the electrons, neutrons, and protons.

Attraction—when two charges or poles are different they "attract " or come together.

Battery—an object that produces an electric charge by means of a chemical reaction.

Capacitor—an electric component that stores an electrical charge.

Cathode—an electrode that is negatively charged.

Circuit—a closed path for electron flow.

Component—the part or piece of a circuit.

Conductor—material that allows the free flow of electrons, creating electric current.

Conclusion—the outcome of an investigation.

Control—a standard measure of comparison in an experiment. The control always stays constant.

Current Electricity—the continuous flow of electrons through a conductor.

Direct Current—the electric current flows in only one direction around the circuit.

Electricity—the flow of electrons through a conductor.

Electric Meter—an object that measures electric current.

Electrode—a solid part of the battery, usually made from materials such as carbon and zinc.

Electromagnetism—an area of physics that studies the relationship between electricity and magnetism.

Electromagnet—a temporary magnet formed when an electric current flows through a wire coil. The coil is usually wrapped around an iron bar.

Electron—a particle inside an atom that carries a negative electric charge.

Electrostatic Attraction—the attraction that opposite electrical charges have for each other.

Experiment—a means of proving or disproving an hypothesis.

Geographic North—the earth's geographic north pole.

Hypothesis (hi-POTH-e-sis)—an educated guess to a question which one is trying to answer.

Insulator—material that will not allow electric current to flow through it.

Investigation—observation of something followed by a systematic inquiry in order to understand what was originally observed.

Magnet—an object that has a magnetic field around it.

Magnetic Field—the area around a magnet that causes magnetic movement.

Magnetic North—the earth's magnetic north pole. This pole continually changes with the earth's magnetic field.

Magnetic Pole—the ends of a magnet. One pole is north, and one pole is south.

Glossary *(cont.)*

N O P Q R S T V

Magnetism—an invisible force that can make objects move away, move together, or stay in the same place.

Neutron—a particle inside an atom that carries a neutral charge.

Observation—careful notice or examination of something.

Procedure—a series of steps that is carried out when doing an experiment.

Proton—a particle inside an atom that carries a positive electrical charge.

Question—a formal way of inquiring about a particular topic.

Repel—to push away. When two charges or poles are the same, they *repel* each other.

Resistance—opposition to the flow of electrons.

Results—the data collected after performing an experiment.

Rheostat—an object that regulates electrical current.

Scientific Method—a creative and systematic process of proving or disproving a given question, following an observation. Observation, question, hypothesis, procedure, results, conclusion, and future investigations comprise the scientific method.

Scientific-Process Skills—the skills necessary to have in order to be able to think critically. Process skills include: observing, communicating, comparing, ordering, categorizing, relating, inferring, and applying.

Scientist—a person considered an expert in one or more areas of science.

Semi-Conductor—a material that conducts electricity better than insulators, but not as well as such conductors as copper.

Static Electricity—electrons that do not move but still create an electrical charge.

Transformer—a device that changes the voltage of electricity.

Variable—the changing factor of an experiment.

Volt—the unit of force by which electricity is measured.

Voltage—a force that causes electrons to flow.

Bibliography

Ardley, Neil. ***Discovering Electricity.*** Watts LB, 1984.

Ardley, Neil. ***Science Book of Electricity.*** HBJ, 1991 .

Ardley, Neil. ***Science Book of Magnets.*** HBJ, 1991 .

Bailey, Mark. ***Electricity***. Raintree Steck-V, 1988.

Baines, Rae. ***Discovering Electricity***. Troll LB, 1982.

Baker & Haslam Staff. ***Electricity***. Macmillan Child Group, 1993.

Berger, Melvin. ***Switch On, Switch Off.*** Harper Collins, 1989.

Brandt, Keith. ***Electricity***. Troll Assocs., 1985.

Burton, Virginia Lee. ***Mike Mulligan & His Steam Shovel.*** Houghton & Mifflin, 1939.

Challand, Helen. ***Experiments with Magnets***. Childrens Press, 1986.

Clemence, John. ***Electricity***. Garrett Ed. Grp., 1991.

Cooper, Jason. ***Magnetism***. Rourke Corp., 1992.

Davis, Kay & Wendy Oldsfield. ***Electricity & Magnetism.*** Raintree Steck-V, 1991.

Dempsey, Michael & Angela Sheehan. ***Water***. World, 1970.

Disch, Thomas. ***The Brave Little Toaster.*** Doubleday, 1986.

Dunn, Andrew. ***It's Electric***. Thompson Learning, 1992.

Gosnell, Kathee. ***Electricity***. Teacher Created Materials, 1994.

Guthridge, Sue. ***Thomas A. Edison: Young Inventor.*** Macmillan C Group, 1986.

Henry, Lucia K. ***Science & Ourselves***. Fearon Teach Aids, 1989.

Hoyt, Marie A. ***Magnet Magic Etc.*** Educ Serv Pr., 1983.

Jennings, Terry. ***Electricity***. Watts, 1990.

Jennings, Terry. ***Electricity & Magnetism.*** Childrens, 1989.

Kaufman, Mervyn. ***Thomas Alva Edison: Miracle Worker.*** Chelsea Hse, 1993.

Kirkpatrick, Rena. ***Look at Magnets***. Raintree LB, 1978.

Lawson, Robert. ***Ben & Me.*** Little Brown, 1988.

Parker, Steve. ***Thomas Edison & Electricity.*** Harp C Child Books, 1992.

Santrey, Laurence. ***Magnets***. Troll Assocs., 1985.

Simon, Seymour. ***Einstein Anderson, Science Sleuth.*** Puffin, 1980.

Vivian, Charles. ***Science Experiments & Amusements for Children.*** Dover, 1967.

Ward, Allen. ***Magnets & Electricity***. Watts, 1992.

Webster, Vera. ***Science Experiments***. Childrens Press, 1982.

Weinberg, Michael. ***Thomas Edison***. Longmeadow Press, 1988.

Whyman, Kathryn. ***Electricity & Magnetism.*** Gloucester, 1986.

Whyman, Kathryn. ***Sparks to Power Stations.*** Gloucester LB, 1989.

Bibliography *(cont.)*

Electricity. Price Stern, 1988.

Spanish Titles

Taylor, K. *Luz (Light)*. Lectorum, 1992.

Technology

Agency for Instructional Technology. ***Magnetism: Why Does a Compass Point North?, Electricity: Where Does Electricity Come From?, Let's Explore Magnets.*** Available from ATI The Learning Source, (800)457-4509. Video Program

Bill Walker Productions. ***Electricity and Magnets.*** Available from Cornet/MTI Film & Video, (800)777-8100. (available in Spanish/English) Film, Video, and Videodisc

Disney Educational Productions. ***Electricity***. Available from Cornet/MTI Film & Video, (800)777-8100. Film and Video

Science Books & Films American Association for The Advancement of Science. ***About Electricity.*** Available from AIMS Media, (800)367-2467. Videodisc

